General Physics 2

PHY2049L Lab Manual
Fifth Edition

De Huai Chen

Shen Li Qiu

Physics Department
Florida Atlantic University

All Images provided by De Huai Chen and Shen Li Qiu

Cover image provided by De Huai Chen and Shen Li Qiu

Kendall Hunt
publishing company

www.kendallhunt.com
Send all inquiries to:
4050 Westmark Drive
Dubuque, IA 52004-1840

CONTACT INFORMATION

Dr. De Huai Chen

Laboratory Director of Physics

Florida Atlantic University

Office:	PS129
Phone:	(561) 297 - 1088
Fax:	(561) 297 - 2662
Email:	dchen@fau.edu
Webpage:	http://wise.fau.edu/~dchen

General Physics 2 Lab --- PHY2049L

Table of Contents

Introduction

PHY2049L is a one-credit lab course and requires two hours of laboratory participation per week. Its experiments address topics within the field of "Electricity, Magnetism and Optics", and thus serve as an experimental counterpoint to topics discussed in the lecture courses including PHY2044, PHY2054, PHY2049. Many topics will be treated earlier in lab syllabus than in lecture, therefore it is essential to read assigned lab manuals before coming to the lab.

The main purpose of physics laboratory is to provide "hands-on" experiences of various physical principles. Theory of a physical principle will be introduced for each experiment, and the predicted results from theory will be tested by experimental measurements. You will learn in this course how a problem in experimental physics is tackled: how to organize the investigation, collect data, analyze the data, draw conclusions and present the results in a written form.

Course Policies for General Physics Labs:

1. **Final grades**

   Final grades are calculated as 80% of the average lab reports and 20% of the average quizzes.

2. **Total number of labs in a semester**

   (a) There are 12 labs in spring and fall semesters. Students can drop a lab with the lowest grade.

   (b) There are 10 labs in summer A and B semesters. No lab-drop is allowed.

3. **Lab reports**

   (a) **Grading scale of lab reports:**

		A	19.0---20	A-	18.0---19.0
B+	17.0---18.0	B	16.0---17.0	B-	15.0---16.0
C+	14.0---15.0	C	13.0---14.0	C-	12.0---13.0
D+	11.0---12.0	D	10.0---11.0	D-	9.0 ---10.0
F	< 9.0				

   (b) Students are expected to write lab reports individually from beginning to end. Lab reports copied from others are unacceptable.

   (c) Students can tear those pages out of the lab manual as a part of their lab reports, which contain measured (raw) data and analyzed data, answers to questions. The data sheets must be checked and signed by the lab TA.

   (d) Each lab report should be turned in one week after the experiment is done. Points will be deducted for late lab reports in the following manner: One week late - one point. Two weeks late - three points. Three weeks late - ten points.

   (e) Lab will be closed one week after the last experiment is done. All lab reports must be turned in by the deadline listed just below Table 1 in the syllabus of PHY2049L.

   (f) Lab reports should include the following contents:
   - Title of experiment, lab date and report date, your name, lab section and lab instructor.
   - The purpose of the experiment.
   - A short summary of the theory underlying the experiment.
   - Experimental method and apparatus including circuit diagram in some experiments.
   - Presentation of your results including raw data and analyzed data in the forms of tables, graphs, photos, fitting plots, etc. The data sheets with TA's signature should be a part of the presentation.
   - Comparison of experimental results with theoretical predictions, and error analysis.
   - Discussions and conclusions.
   - Answers to the questions listed in the lab manual of each experiment.

4. **Quizzes**
   (a) In spring and fall semesters students take 4 close-book quizzes, and can drop a quiz (except the hand-on quiz) with the lowest grade. In summer A and B students take 3 close-book quizzes without dropping any quiz. The final quiz score is the average of three quizzes. The quizzes are meant to check students' understanding of physics concepts of previous lab reports, which might be either written or hand-on quizzes.
   (b) No makeup quiz is allowed unless you have a written document such as a letter from your doctor showing that you are unable to come to the lab for taking the quiz.

5. **Lab performance and maintenance**
   (a) Students are expected to go through each lab section in the lab manual before class.
   (b) Students should conduct the non-computerized experiments individually.
   (c) Students should conduct the computerized experiments in a group and share the collected data. Students should follow the instructions given in the lab manual and do the preparation first, making sure that the lab setup is correct before collecting data.
   (d) Before leaving the lab, students should let TA check their measured (raw) data and analyzed data (such as data plots and fittings in capstone files). If all the data are correct, TA will sign the data sheets. If mistakes or unacceptable large errors are found in the raw data, TA will ask students redo some measurements to get all the raw data correct.
   (e) If time permits, students are strongly encouraged to complete all required analysis and calculations for the experiment before leaving the lab.
   (f) Students by no means should change computer settings, or use computers for any other purposes such as typing or printing lab reports.
   (g) Students must report to their instructor any damage or loss of equipment.
   (h) No food is allowed in the lab. Before leaving the lab students should turn off power of electronic devices, put lab instruments in order, take away drinking water bottle, paper towels, etc.

6. **Lab makeup policy**
   (a) The lab equipment is changed once a week in spring and fall semesters, twice a week in summer semesters. Students are allowed to makeup a lab before the lab equipment is changed. Students should get permission from their instructor to makeup a lab in another lab section. The lab report should be turned in to the instructor of the original lab section along with an approval written by the instructor of the lab section in which the makeup lab is done.
   (b) To makeup a lab after the lab equipment is changed students must have a good reason such as illness (with doctor's letter), unpredicted event such as car accident (with police report), etc.

Computerized Experiment

Many experiments in PHY2049L are computerized using PASCO 850 universal interface and Capstone software.

1. PASCO 850 Universal Interface

PASCO® 850 Universal Interface is a USB (Universal Serial Bus) multi-port data acquisition interface designed for use with any PASCO sensor and PASCO Capstone™ software. Users can plug a sensor into one of the twelve input ports on the interface, perform the necessary setup in the PASCO Capstone program, and then begin collecting data. PASCO Capstone software records, displays and analyzes the data measured by the sensor.

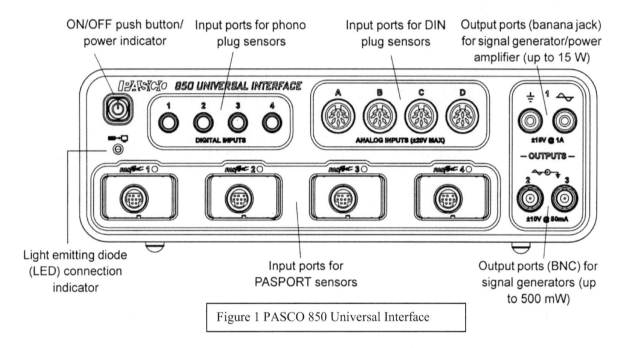

Figure 1 PASCO 850 Universal Interface

The 850 Universal Interface comes with a USB cable for connecting to a computer, and a power supply (AC adapter with power cord) that converts input of 100 to 240 V AC to output of 20 V DC at 6 A.

The 850 Universal Interface has three built-in signal generator/power outputs. One provides up to 15 watts of power and the other two provide 500 milliwatts of power each. The interface can output direct current (DC) or alternating current (AC) in a variety of waveforms such as sine, square, and sawtooth.

2. PASCO Capstone Software

PASCO Capstone software is required for the 850 Universal Interface,
Setup PASCO 850 interface and capstone software:
(a) With PASCO 850 Interface power off make its connections to computer and to sensors required for each experiment.
(b) Turn on PASCO 850 Interface and computer.
(c) Log in computer using password PHY2048L for lab 1 or PHY2049L for lab 2.
(d) Start PASCO Capstone by double clicking the symbol on computer.

The Workbook page appears as shown in Fig. 2.

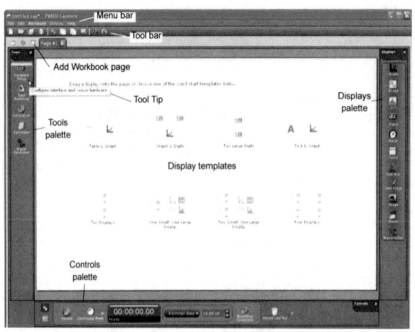

Figure 2 Workbook Page from Capstone
Hover the cursor over an icon, button, or other element to see a descriptive Tool Tip.

More information on PASCO 850 Universal Interface and Capstone software can be found at:
http://www.pasco.com/prodCatalog/UI/UI-5000_850-universal-interface/
http://www.conatex.com/mediapool/betriebsanleitungen/BLE_1124080.pdf
http://www.pasco.com/capstone/

Fundamental quantities and units

A fundamental (base) quantity is one of a conventionally chosen subset of physical quantities, where no subset quantity can be expressed in terms of the others. Fundamental (base) unit is a unit adopted for measurement of a fundamental quantity. In the International System (SI) of Units, there are seven fundamental quantities and units which are listed in the Table.

Fundamental quantity	Symbol	Fundamental SI unit	Symbol
Length	ℓ	meter	m
Mass	m	kilogram	kg
Time	t	second	s
Temperature	T	kelvin	K
Electric current	I	Ampere	A
Amount of substance	n	mole	mol
Luminous intensity	I_{υ}	candelas	cd

Derived quantities and units

All other physical quantities are called derived quantities which can be derived from the fundamental quantities using various physics laws. For instance, speed: $\upsilon = \ell / t$, its unit is m/s; force: $F = ma$, its unit is Newton (N), $1N = 1kg\,m/s^2$, etc.

Experimental error and data analysis

Purpose

To review the types of experimental errors and some methods of error and data analysis that will be used in subsequent experiments.

Theory

Any measurement of physical quantities always involves some uncertainty or experimental error. One should not only report a result but also give some indication of its uncertainty of the experimental data.

1. **Types of experimental errors**
    (a) **Random or statistical errors** result from unknown and unpredictable variation that arise in all experiment situations, for example, fluctuation in temperature or voltage, mechanical vibrations of an experimental setup, unbiased estimates of measurement readings by the observer. Repeated measurements with random error give slightly different values each time. The random error can be estimated by repeating an experiment several times.
    (b) **Systematic errors** are associated with particular measurement instruments or techniques, for example, an improperly calibrated instrument or personal error, such as using a wrong constant in a calculation or always taking a low reading of a scale division. Avoiding systematic errors depends on the skill of the observer to recognize the sources of such error and to prevent or correct them.

2. **Accuracy and precision**
    Accuracy and precision are commonly used synonymously, but in experimental measurements there is an important distinction.
    (a) **Accuracy** signifies how close the measured value comes to the true or accepted value, that is, how correct it is.
    (b) **Precision** refers to the agreement among repeated measurements or how close they are together.

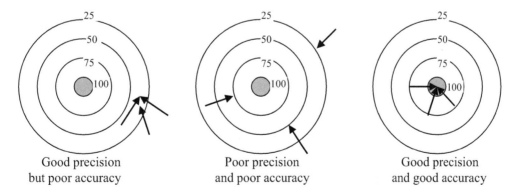

| Good precision but poor accuracy | Poor precision and poor accuracy | Good precision and good accuracy |

3. **Significant figures**
    The significant figures of an experimentally measured value include all the numbers that can be read directly from the instrument scale plus one doubtful or estimated number (fraction of the smallest division). For example, the length of an object may be read as 2.54 cm (three significant numbers) on one instrument scale and 2.5405 cm (five significant numbers) on another. Thus, the significant numbers depend on the quality of the instrument and the fineness of its measuring scale.

4. Computations with measured values

(a) Addition and subtraction: Begin with the first column from the left that contains a doubtful figure, round off all numbers to this column and drop all digits to the right.

$$42.31 \; 0.0621 + 512.4 + 2.57 = 42.3 + 0.1 + 512.4 + 2.6 = 557.4$$

(b) Multiplication and division: The number of significant figures in the final answer in general equals the number of significant figures in the measurement with the least number of significant figures.

$$6.27 * 5.5 = 34.485 \text{ (from a hand calculator)} = 34.$$

$$\frac{374}{29} = 12.896551 \text{ (from a hand calculator)} = 13.$$

5. Expressing experimental error and uncertainty

(a) Percentage error

The **accepted** or "true" value of a physical quantity found in textbooks and handbooks is the most accurate value through sophisticated experiments or mathematical methods.

The **absolute difference** between an experimental value $\bar{x}$ (defined in 5(b)) and the accepted value A, is written in $|\bar{x} - A|$. The percentage error of the experimental value is:

$$\text{Percentage error} = \frac{absolute\ difference}{accepted\ value} * 100\% = \frac{|\bar{x} - A|}{A} \times 100\%$$

The accuracy of the experimental value is expressed in terms of percentage error.

(b) Average (Mean) value ($\bar{x}$)

The average or mean value of a set of N measurements is:

$$\bar{x} = \frac{x_1 + x_2 + x_3 + \ldots + x_N}{N} = \frac{1}{N} \sum_{i=1}^{N} x_i$$

(c) Deviation from the mean

Having obtained a set of measurements and determined the mean value, it is helpful to report how widely the individual measurements are scattered from the mean. A quantitative description of this scatter or dispersion of measurements will give an idea of the precision of the experiment.

- **Deviation** d_i

   The **deviation** d_i from the mean of any measurement with a mean value $\bar{x}$ is
   $d_i = x_i - \bar{x}$.

- **Mean deviation** $\bar{d}$

   The **mean deviation** $\bar{d}$ **is a measure of the dispersion of experimental measurements about the mean, i.e., a measure of precision.**

   To find the **mean deviation** $\bar{d}$ of a set of N measurements, the absolute deviations $|d_i|$ are determined first,

$$|d_i| = |x_i - \bar{x}|$$

The **mean deviation** $\overline{d}$ is then

$$\overline{d} = \frac{|d_1| + |d_2| + |d_3| + \ldots + |d_N|}{N} = \frac{1}{N} \sum_{i=1}^{N} |d_i|$$

Example: What is the average $\overline{x}$ or mean value of the set of numbers 5.42, 6.18, 5.70, 6.01, and 6.32? What is its mean deviation $\overline{d}$?

$$\overline{x} = \frac{1}{N} \sum_{i=1}^{N} x_i = \frac{5.42 + 6.18 + 5.70 + 6.01 + 6.32}{5} = 5.93$$

$$|d_1| = |5.42 - 5.93| = 0.51$$
$$|d_2| = |6.18 - 5.93| = 0.25$$
$$|d_3| = |5.70 - 5.93| = 0.23$$
$$|d_4| = |6.01 - 5.93| = 0.08$$
$$|d_5| = |6.32 - 5.93| = 0.39$$

$$\overline{d} = \frac{1}{N} \sum_{i=1}^{N} |d_i| = \frac{0.51 + 0.25 + 0.23 + 0.08 + 0.39}{5} = 0.29$$

It is common to report the experimental value of a quantity in the form: $\overline{x} \pm \overline{d}$. In the above example, this would be 5.93 $\pm$ 0.29 or 5.9 $\pm$ 0.3.

An experimental value and its mean deviation should have their last significant digits in the same location relative to the decimal point. The $\pm$ gives a measure of the precision of the experimental value.

It is also common practice to express the dispersion on the mean deviation as a percent of the mean:

$$\overline{x} \pm \frac{\overline{d}}{\overline{x}} \times 100\%$$

For the above example, we have 5.93 $\pm \dfrac{0.29}{5.93} \times 100\% = 5.93 \pm 4.9\%$

(d) Standard deviation (optional for PHY 2049L)

Standard deviation, δ, is defined as: $\quad \delta = \sqrt{\dfrac{1}{n-1} \sum_{i=1}^{n} [x_i - \overline{x}]^2}$

(e) Propagation of errors (optional for PHY 2049L)
  • **Addition and subtraction of experimental values**
   Suppose x, y, and z are three measured values and the errors are $\delta_x, \delta_y, \delta_z$. If w is the value to be calculated from these measurements and is defined to be:
   $$w = x - y + z$$
   Then statistical analysis shows that in good approximation the error δ_w is:
   $$\delta_w = \sqrt{\delta_x^2 + \delta_y^2 + \delta_z^2}$$

  • **Multiplication and division of experimental values**
   Suppose x and y are two measured values and the errors are δ_x and δ_y. If w is the value to be calculated from these measurements and is defined to be:

$$w = x * y \ (\text{or } w = \frac{x}{y})$$

Then statistical analysis shows that the fractional error of w is:

$$\frac{\delta_w}{w} = \sqrt{\left(\frac{\delta_x}{x}\right)^2 + \left(\frac{\delta_y}{y}\right)^2}$$

6. Graphical representation of data

It is often to represent experimental data in graphical form, not only for reporting, but also to obtain information.

Quantities are commonly plotted using rectangular Cartesian axes (x and y). The location of a point on the graph is defined by its coordinates x and y, referenced to the origin O.

When plotting data, choose axis scale so that most of the graph paper is used and the scale units should always be included.

When the data points are plotted, draw a smooth line with an approximately equal number of points on each side to make a "curve of best fit".

In case where several determinations of each experimental quantity are made, the average value is plotted and the mean deviation may be plotted as error bars.

Figure 1 shows an example of a properly labeled and plotted graph with error bars indicating mean deviation.

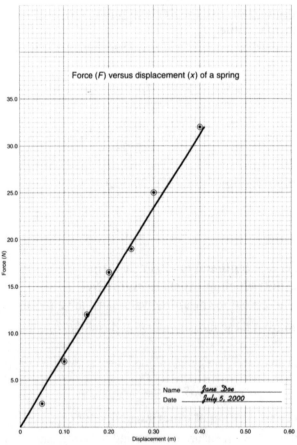

Figure 1 Proper graphing. An example of a properly labeled and
plotted graph, see text for description.

Experiment 1
Electric field and electric potential

Purpose
Map electric equipotential lines and electric field lines for two-dimensional charge configurations.

Equipment
Three sheets of conductive papers with conductive-ink electrodes, push pins and wires, a comfort grip voltage probe with silicone insulated, marker pen, ruler, (15 cm), corkboard, digital multimeters (DMM), DC power supply (HY152A).

Theory
The magnitude of electrostatic force between two point charges Q and q is given by Coulomb's law:

$$F = k \frac{|Qq|}{r^2} \qquad (1)$$

where r is the distance between the charges, k is the Coulomb constant ($k = 9.0 \times 10^9$ Nm^2/C^2). The directions of the forces the two charges exert each other are always along the line joining the two point charges and like charges repel, unlike charges attract.

The magnitude of the electric field is defined as the electrical force per unit charge, i.e., $E = F/q$ (N/C), here q is a test point charge. In the case of the electric field associated with a single-source charge Q, the magnitude of the electric field a distance r away from the source charge is

$$E = \frac{F}{q} = k \frac{|Q|}{r^2} \qquad (2)$$

Electric field is a vector quantity which is specified by both its magnitude and direction. Using the unit vector $\hat{r} = \dfrac{\vec{r}}{r}$, we can write a vector equation that gives both the magnitude and direction of the electric field $\vec{E}$ due to a point charge Q:

$$\vec{E} = k \frac{Q}{r^2} \hat{r} \qquad (3)$$

The electric field of a point charge always points away from a positive charge but towards a negative charge.

Figure 1 (a) shows the electric field vectors from a positive source charge. By drawing lines through the points in the direction of the field vectors, we form the electric field lines (Fig. 1 b & c), which give a graphical representation of the electric field.

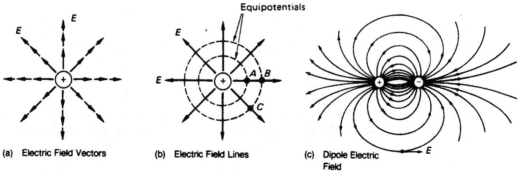

(a) Electric Field Vectors   (b) Electric Field Lines   (c) Dipole Electric Field

Figure   1

It can be shown (see your textbook) that the electric potential at a distance r from the source charge Q is:

$$V = k\frac{Q}{r}.\tag{4}$$

Electric potential is a scalar quantity which is specified by its magnitude only. Electric potential can be either positive or negative depends on the sign of the source charge Q.

The electric potential of two or more point charges is simply the algebraic sum of the potentials due to each point charge separately:

$$V = k\sum_{i=1}^{N}\frac{Q_i}{r_i}, \text{ where N is the total number of point charges.}\tag{5}$$

Since a free charge moves in an electric field by the action of the electric force, we say that work (W_{AB}) is done by the field on moving a charge from one point to another (e.g. from A to B in Fig. 1 b). To move a positive charge from B to A would require work supplied by an external force to move the charge against the electric field. The work done by an electric field in moving a unit charge from point A to point B is called the potential difference between these two points:

$$\Delta V_{AB} = V_A - V_B = \frac{W_{AB}}{q}\tag{6}$$

If a charge is moved along a path perpendicular to the field lines, no work is done, i.e., W = 0, since there is no force component along the path. Then along such a path (Fig. 1b between B and C) $\Delta V_{BC} = V_B - V_C = \frac{W_{BC}}{q} = 0$, i.e., $V_B = V_C$. Hence the potential has a constant value along a path perpendicular to the field lines. Such a path is called an equipotential line which is always perpendicular to the field lines. Different equipotential lines have different potential values, therefore, equipotential lines can never intercept.

An electric field may be mapped experimentally by determining either the field lines or the equipotential lines. Equipotential line can be determined by measuring the potential difference ΔV between two points. If $\Delta V = 0$, the two points are on an equipotential line.

Caution: Make sure that the "+" and "−" leads from the DC power do not touch each other before you turn on the power supply in all the measurements.

Measurement #1: Map equipotential lines for parallel-plates capacitor.
1. Mount the conductive paper #1 (with printed side up) on the corkboard using two metallic pushpins at the two electrodes which are drawn with conductive ink on the conductive paper #1.
2. Connect the two electrodes on the conductive paper #1 to the DC power supply (HY152A).
3. Connect the black banana jack (ground) on DMM to the ground (black banana jack) on the DC power supply (HY152A). Connect the red banana jack on DMM to a comfort grip voltage probe with silicone insulated. Select voltage scale of 20 V on DMM.
4. Turn on the DC power to 10 V.
5. To measure the voltage at a point on the conductive paper #1 in Fig. 2, just let the tip of the voltage probe touch that point. The tip of the voltage probe should be perpendicular to the conductive paper at the touch point (in Fig. 2).
6. **Attention:** To reduce measurement error, at each labelled cross in Fig. 2, say 1A, **measure three voltages at three positions along the vertical line of the cross**. Make the same "3-position-voltage" measurements at each labelled cross (from 1A to 5E) in Fig. 2. Record the data in Table 1.
7. Turn the voltage to 0V before shut down the DC power (HY152A).

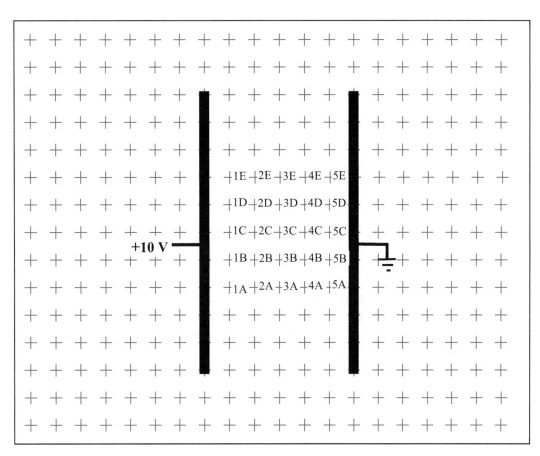

Figure 2   Conductive paper #1

Table 1   Data of 3-position-voltage measurements at each labeled cross (from 1A to 5E) in Fig. 2.

	Voltage (V)														
	A			B			C			D			E		
1															
2															
3															
4															
5															
Ave.															

Data Analysis 1

1. Draw equipotential lines on Fig. 2 using a red pen. Draw field lines (from high to low voltage) using a blue pen.
2. Are the field lines perpendicular to the equipotential lines?

3. Use a ruler to measure Δx on the conductive paper. Record Δx value in Table 2.
4. Calculate the magnitude of the electric field E for the parallel plate configuration using formula $E = \Delta V / \Delta x$. Record the data in Table 2.

Table 2 Potential difference between two vertical lines with indices 1, 2, ….5 labeled in Fig. 2.

ΔV	$\Delta V_{12} = V_1 - V_2$	$\Delta V_{23} = V_2 - V_3$	$\Delta V_{34} = V_3 - V_4$	$\Delta V_{45} = V_4 - V_5$
$\Delta x =$				
$E = \Delta V / \Delta x$				

5. Are the E values close each other?

6. Is it true that $\vec{E}_{1A} = \vec{E}_{1B} = \ldots\ldots = \vec{E}_{5E}$? Why?

Measurement #2:   Map equipotential lines for a point charge with guard rings.

1. Mount the conductive paper #2 (with printed side up) on the corkboard using two metallic pushpins at the two electrodes which are drawn with conductive ink on the conductive paper #2.
2. Connect the two electrodes on the conductive paper #2 to the DC power supply (HY152A).
3. Connect the black banana jack (ground) on DMM to the ground (black banana jack) on the DC power supply (HY152A). Connect the red banana jack on DMM to a comfort grip voltage probe with silicone insulated. Select voltage scale of 20 V on DMM.

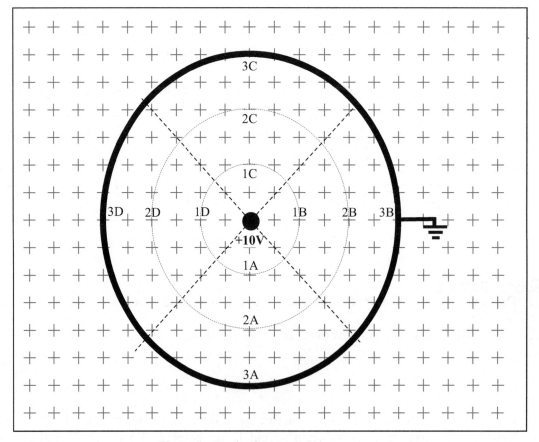

Figure 3   Conductive paper #2

4. Turn on the DC power to 10 V.
5. **Attention:** To reduce measurement error, the tip of the voltage probe should be perpendicular to the conductive paper at the touch point.
6. Measure voltages at all the points labeled in Fig. 2. Record the data in Table 3.
7. Turn the voltage to 0V before shut down the DC power (HY152A).

Table 3   Measured voltages at all the points labeled in Fig. 3 (conductive paper #2).

	Voltage (V)			
	A	B	C	D
1				
2				
3				

Data Analysis 2

1. Based on the data in Table 3 draw equipotential lines on Fig. 3 using a red pen. Draw field lines (from high to low voltage) using a blue pen. Are the field lines perpendicular to the equipotential lines?

2. Theoretically, should the points 1A to 1D be on an equipotential line? Are the measured voltage values at points 1A to 1D very close each other? If not, what is the main source of error?

Measurement #3: Map equipotential lines for two-point-charge configuration (at +10V and 0 V respectively).

1. Mount the conductive paper #3 (with printed side up) on the corkboard using two metallic pushpins at the two electrodes which are drawn with conductive ink on the conductive paper #3.
2. Connect the two electrodes on the conductive paper #3 to the DC power supply (HY152A).
3. Connect the black banana jack (ground) on DMM to the ground (black banana jack) on the DC power supply (HY152A). Connect the red banana jack on DMM to a comfort grip voltage probe with silicone insulated. Select voltage scale of 20 V on DMM.
4. Turn on the DC power to 10 V.
5. Attention: To reduce measurement error, the tip of the voltage probe should be perpendicular to the conductive paper at the touch point.
6. Measure voltages at all the points labeled in Fig. 4. Record the data in Table 4.
7. Turn the voltage to 0V before shut down the DC power (HY152A).

Table 4   Measured voltages at all the points labeled in Fig. 4.

	Voltage (V)				
	A	B	C	D	E
1					-------
2					-------
3					
4					-------
5					-------

Data Analysis 3

1. Based on the data in Table 4 draw equipotential lines on Fig. 4 using a red pen.

2. What is the main cause for the difference in shape of the equipotential lines between Fig. 3 and Fig. 4?

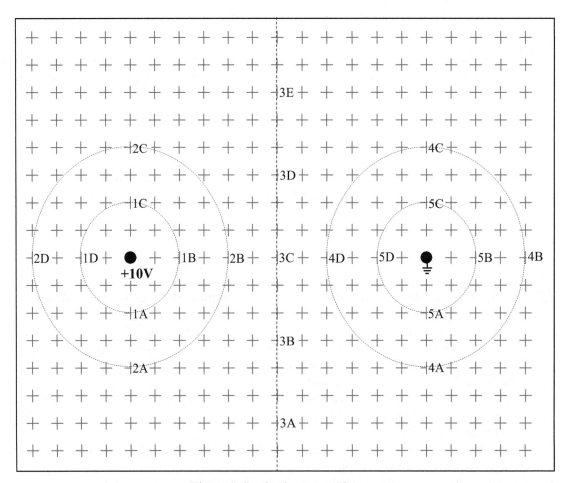

Figure 4 Conductive paper #3

Work to be done:

1. Check that the DC power (HY152A) is off. If not, turn voltage to 0V then shut it down.
2. Let your TA check your data pages. If they are OK, your TA will sign them.
3. Clean up your bench.

Lab report on Experiments 1

1. Your lab report should be in the required format described in the "Introduction" of the lab manual.
2. Figures 2, 3 and 4 with recorded data should be included in your lab report.
3. Tables 1 to 4 should be included in your lab report.
4. It is required that the answers to the questions listed in each data analysis and in "Questions and Exercises" should be included in your lab report.
5. You can tear those pages out of the lab manual as a part of your lab report, which contain measured (raw) data and analyzed data, answers to questions. The data sheets must be checked and signed by your lab TA.

Questions and Exercises

The following three electrostatic charge configurations are in three separated spaces.

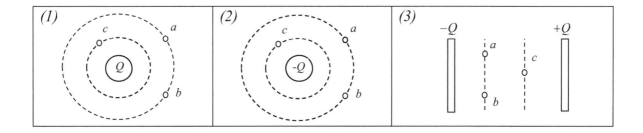

1. Use ruler to draw vector arrows with label $\vec{E}_a, \vec{E}_b, \vec{E}_c$ on each diagram to indicate the electric field lines through points a, b, c respectively in each of the three charge configurations.

2. Let E_a, E_b, E_c be the magnitudes of $\vec{E}_a, \vec{E}_b, \vec{E}_c$ respectively. Use smaller (<) , greater (>) and equal (=) signs to describe the relationships among E_a, E_b, E_c in the configurations:

      configuration (1): ._____.

      configuration (2): ._____.

      configuration (3): ._____.

3. The electric potentials at points a, b, c are denoted V_a, V_b and V_c respectively. Use smaller (<) , greater (>) and equal (=) signs to describe the relationships among V_a, V_b and V_c in:

configuration (1): _____.

configuration (2): _____.

configuration (3): _____.

4. Now a negative point charge $-q$ ($|-q| \ll Q$) is placed at point a.

Use ruler to draw a vector arrow (with a label $\vec{F}_a$) on each diagram to indicate the direction of the motion of q in each of the three charge configurations.

Experiment 2
Resistors in series and parallel connections (1)

Purpose
1. Practice on the use of digital multimeter (DMM).
2. Light bulbs in series and parallel connections.

Equipment
One digital multimeter (DMM), DC power supply #1 (5 V adapter), DC power supply #2 (HY152A), four light bulbs with sockets, banana connectors. Demo box of resistors.

Theory
When a voltage or potential difference (V) is applied across a conductor, the current (I) in the conductor is found to be proportional to the voltage. The resistance (R) of the conductor is defined as the ratio of the applied voltage and the resulting current, that is,

$$R = \frac{V}{I} \tag{1}$$

A conductor or an electrical device offers resistance to the current, it is called a resistor.
Resistors are connected in series if they are connected in "head to tail" as shown in the drawing.

$$R_1 \qquad R_2$$

By the conservation of charges, the current flows through each resistor is the same, $I = I_1 = I_2$, but the sum of the voltage drops is $V = V_1 + V_2 = I R_1 + I R_2 = I (R_1 + R_2) = I R_s$. R_s is the equivalent resistance of R_1 and R_2 in series connection.

$$R_s = R_1 + R_2 \tag{2}$$

Resistors are connected in parallel if they are connected in "head to head and tail to tail" as shown below.

The voltage drops across all the resistors are the same and equal to the voltage V of the source, $V = V_1 = V_2$, but the current from the source divides among the resistors, such that

$$I = I_1 + I_2 = \frac{V}{R_1} + \frac{V}{R_2} = \frac{V}{R_p} \qquad\qquad \frac{1}{R_p} = \frac{1}{R_1} + \frac{1}{R_2}$$

$$R_p = \frac{R_1 R_2}{R_1 + R_2} \tag{3}$$

R_p is the equivalent resistance of R_1 and R_2 in parallel connection. The change in electric energy per unit time is called electric power. The power dissipated in a resistor is given by

$$P = IV = I^2 R \tag{4}$$

The R in (4) can be the resistance of a single resistor, or the equivalent resistance R_s or R_p.

DMM ----- Digital multimeter

DMM (digital multimeter) is a measuring instrument which combines **Ammeter, Voltmeter** and **Ohmmeter** into a single instrument. An Ammeter measures current, a Voltmeter measures the potential difference (voltage) between two points, and an Ohmmeter measures resistance. As an example Fig. 1 shows one of the DMMs used in our lab.

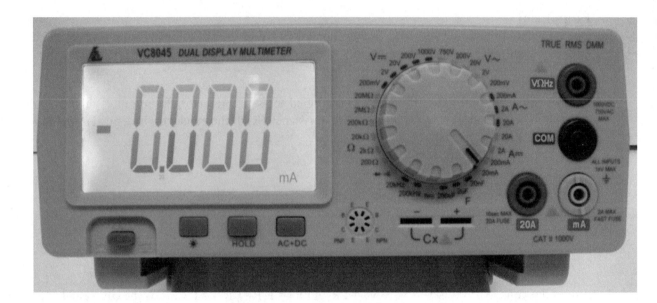

Figure 1   DMM (VC8045) which measures direct current (dc) and alternating current (ac) from 0 to 20 amps (A); ac voltage from 0 to 750 volts (V), dc voltage from 0 to 1000 volts; resistance from 0 to 20 $M\Omega$.

1. Use DMM to measure dc current through the loop in Fig. 2
    Connect the red test lead to "mA" terminal (max. 0.2 A) and the black one to COM. Set the function knob to A⎓ range. Now the DMM becomes an **Ammeter**. If the current to be measured is unknown beforehand, start from the highest scale and work down to select a proper scale for your measurement. Notice that **you need to break the circuit**, e.g., pull out a leg of R_1 from the circuit, then insert the ammeter at the break point so that the R_1 and ammeter are in series connection as shown in Fig. 2.

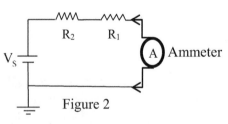

2. Use DMM to measure resistance
    Connect the red test lead to VΩHz terminal and the black one to COM. Set the function knob to Ω range. Now the DMM becomes an **Ohmmeter**. If the resistance to be measured is unknown beforehand, start from the highest scale and work down to select a proper scale for your measurement. Place a resistor on an insulating table, use the two tips of the probes to touch the two legs (electrodes) of the resistor, the value of the resistance is displayed on the LCD (liquid crystal display). **If the resistor to be measured is a part of a circuit you need to break the circuit**, i.e., pull out one leg of the resistor from the circuit then measure its resistance.

3. Use DMM to measure dc voltage across R_1 in Fig. 3
Connect the red test lead to VΩHz terminal and the black one to
COM. Set the function knob to V= range. Now the DMM
becomes a dc **Voltmeter**. If the voltage to be measured is unknown
beforehand, start from the highest scale and work down to select a
proper scale for your measurement. Use the two tips of the probes to
touch the two legs of R_1 to read the voltage across R_1. Vs is the
output voltage of the power supply. Notice that **you don't need to
break the circuit** for measuring the voltage across an element in a
circuit.

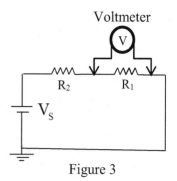

Figure 3

Measurement #1: Light bulbs in series connection

Note: The symbol ⊗ represents a light bulb which can be treated as a resistor.

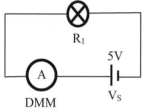

Figure 4 (a)

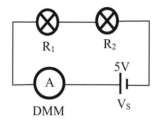

Figure 4 (b)

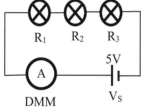

Figure 4 (c)

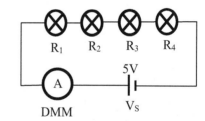

Figure 4 (d)

1. Check that the DC power supply #1 is off. Connect the circuit shown in Fig. 4(a).
2. Turn on the DC power supply #1.
3. Record the DMM reading and the observed bulbs' brightness in Table 1.
4. Turn off the DC power supply #1.
5. Repeat steps 1 to 4 for circuit shown in Fig. 4(b).
6. Repeat steps 1 to 4 for circuit shown in Fig. 4(c).
7. Repeat steps 1 to 4 for circuit shown in Fig. 4(d).
8. Repeat steps 1 to 4 for circuit shown in Fig. 4(d) but with one bulb is removed from its socket.

Table 1 DMM reading and observed bulbs' brightness for the circuits in Fig. 4.

Source voltage	Cases	DMM reading	Current through power supply	Current through each bulb	Observed brightness
$V_S \sim 5V$	Fig. 4(a)				
	Fig. 4(b)				
	Fig. 4(c)				
	Fig. 4(d)				
	Fig. 4(d) but with one light bulb removed				

Questions on Measurement #1:

1. Is the current through the power supply and each bulb (in Table 1) increasing or decreasing with increasing the number of bulbs from Fig. 4 (a) to Fig. 4 (d)? Why?

2. Is there any correlation between the current through each bulb and the observed bulbs' brightness listed in Table 1? If yes, is the correlation consistent with $P = IV = I^2R$?

3. Are the lights at your home connected in series? Why?

Measurement #2: **Light bulbs in parallel connection**

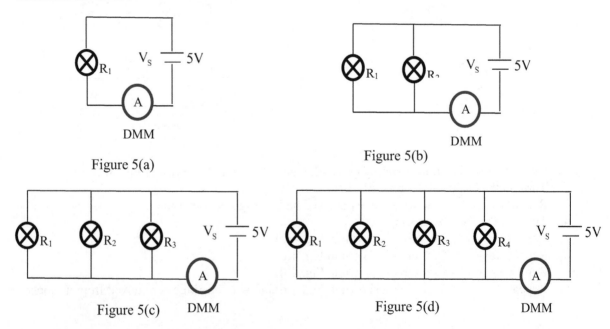

Figure 5(a)

Figure 5(b)

Figure 5(c)   DMM

Figure 5(d)   DMM

1. Check that the DC power supply #1 is off. Connect the circuit shown in Fig. 5(a).
2. Turn on the DC power supply #1.
3. Record the DMM reading and the observed bulbs' brightness in Table 2.

4. Turn off the DC power supply #1.
5. Repeat steps 1 to 4 for circuit shown in Fig. 5(b).
6. Repeat steps 1 to 4 for circuit shown in Fig. 5(c).
7. Repeat steps 1 to 4 for circuit shown in Fig. 5(d).
8. Repeat steps 1 to 4 for circuit shown in Fig. 5(d) but with one bulb is removed from its socket.

Table 2 DMM reading and observed bulbs' brightness for the circuits in Fig. 5.

Source voltage	Cases	DMM reading	Current through power supply	Current through each bulb	Observed Brightness
$V_S \sim 5V$	Fig. 5(a)				
	Fig. 5(b)				
	Fig. 5(c)				
	Fig. 5(d)				
	Fig. 5(d) but with one light bulb removed				

Questions on Measurement #2:
1. Is the current through the power supply and each bulb (in Table 2) increasing or decreasing with increasing the number of bulbs from Fig. 5 (a) to Fig. 5 (d)? Why?

2. Is there any correlation between the current through each bulb and observed bulbs' brightness? Why?

3. Are the lights at your home connected in parallel? Why?

4. The maximum output current from the DC power supply #1 is 2 A. What is the maximum number of light bulbs (used in this experiment) can be connected in parallel with the DC power supply #1? What will happen if **more than** 10 such light bulbs are connected in parallel with the DC power supply #1?

Measurement #3:   **Light bulbs in complex connections**

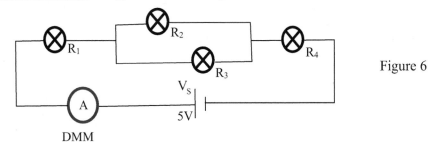

Figure 6

1. Check that the DC power supply #1 is off. Connect the circuit shown in Fig. 6.
2. Turn on the DC power supply #1.
3. Record the measured current and observed bulbs' brightness in Table 3.
4. Turn off the DC power supply #1.

5. Repeat steps 1 to 4 with a change: use DC power supply #2 (HY152A) and set the output voltage to $V_S = 10V$.
6. Repeat step 5 with a change: remove light bulb #1 from its socket and keep the other three light bulbs in their original places.
7. Repeat step 5 with a change: remove light bulb #2 from its socket and keep the other three light bulbs in their original places.
8. Repeat step 5 with a change: remove light bulb #3 from its socket and keep the other three light bulbs in their original places.
9. Repeat step 5 with a change: remove light bulb #4 from its socket and keep the other three light bulbs in their original places.

Table 3 DMM reading and observed bulbs' brightness for the circuits in Fig. 6.

Source voltage	Cases	DMM reading	Observed bulbs' brightness			
			bulb #1	bulb #2	bulb #3	bulb #4
$V_S \sim 5V$	All 4 light bulbs are in circuit					
$V_S = 10V$	All 4 light bulbs are in circuit					
	Light bulb #1 is removed					
	Light bulb #2 is removed					
	Light bulb #3 is removed					
	Light bulb #4 is removed					

Questions on Measurement #3:
1. Although currents I_2 and I_3 are not measured directly you can find the values of I_2 and I_3 from the measured current in Table 3 (assume all the four light bulbs have the same resistance).
   Find the values of I_2 and I_3 (with all the four bulbs in the circuit) at
   (a) $V_S = 5V$: $I_2 =$ _____ (mA), $I_3 =$ _____ (mA)
   (b) $V_S = 10$ V: $I_2 =$ _____ (mA), $I_3 =$ _____ (mA)
   Using these data explain why bulbs #2 and #3 are brighter at $V_S = 10$ V than at $V_S \sim 5$ V.

2. Using above data and the measured data of current in Table 3 explain:
   (a) why bulbs #1 and #4 are brighter than bulbs #2 and #3 at both $V_S \sim 5$ V and 10 V when all the 4 bulbs are in circuit;

   (b) why bulbs #3 becomes brighter after bulb #2 is removed (all other bulbs are remained in the circuit).

3. Can the lights at your home be connected as shown in Fig. 6? Why?

Work to be done:
1. Check that the DC power supply #1 is off.
2. Check that the DC power supply #2 (HY152A) is off. If not, turn voltage to 0V then shut it down.
3. Let your TA check your data Tables. If they are OK, your TA will sign them.
4. Clean up your bench.

Lab report on Experiments 2
1. Your lab report should be in the required format described in the "Introduction" of the lab manual.
2. Tables 1 to 3 along with their corresponding circuits should be included in your lab report.
3. It is required that the answers to the questions on each measurement (listed below Tables 1 to 3) should be included in your lab report.
4. You can tear those pages out of the lab manual as a part of your lab report, which contain measured (raw) data and analyzed data, answers to questions. The data sheets must be checked and signed by your lab TA.

Experiment 3
Resistors in series and parallel connections (2)

Purpose
1. Investigate the relationship between current and voltage in Ohmic materials.
2. Investigate the characteristics of resistors in series and parallel connections.
3. Use DMM (digital multimeter) to measure resistance, dc voltage and dc current.

Equipment
Four resistors, one digital multmeter (DMM), DC power supply (5 V adapter), banana connectors. Demo box of resistors.

Theory   Same as in Experiment 2.

DMM ----- Digital multimeter   Same as in Experiment 2.

Measurement #1: use Ohmmeter to measure resistance
1. Use a Ohmmeter to measure the resistances of R_1, R_2, R_3, R_4 and the equivalent resistances of their combinations, record the measured data in row #2 of Table 1.
   R_a --- the equivalent resistance of R_1 and R_2 in series connection shown in Fig. 1a.
   R_b --- the equivalent resistance of R_2 and R_3 in parallel connection shown in Fig. 1b.
   R_c --- the equivalent resistance of R_1, R_2 and R_3 in the connections shown in Fig. 1c.
   R_d --- the equivalent resistance of R_1, R_2, R_3 and R_4 in series connection shown in Fig. 1d.
   R_e --- the equivalent resistance of R_1, R_2, R_3 and R_4 in the connections shown in Fig. 1e.
   R_f --- the equivalent resistance of R_1, R_2, R_3 and R_4 in the connections shown in Fig. 1f.

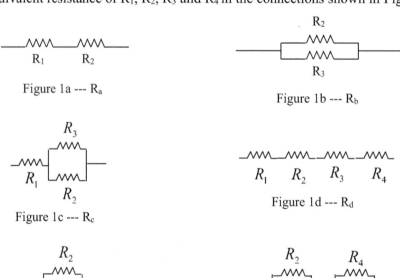

Figure 1a --- R_a

Figure 1b --- R_b

Figure 1c --- R_c

Figure 1d --- R_d

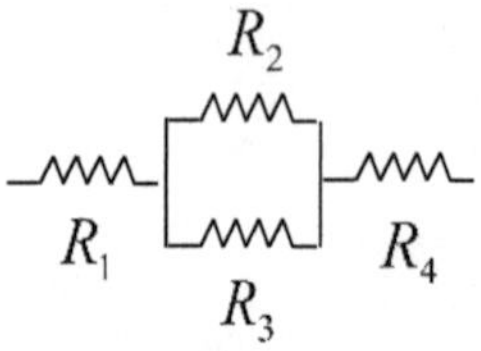

Figure 1e --- R_e

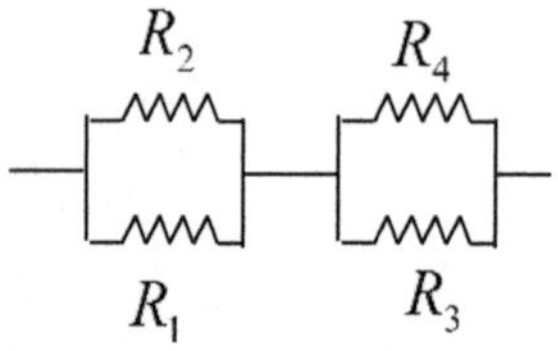

Figure 1f --- R_f

2. Write down formulas for the equivalent resistances of R_a, R_b, R_c, R_d, R_e and R_f in row #3 of Table 1.
3. Calculate R_a, R_b, R_c, R_d, R_e and R_f using the formula in row #3 and record the calculated values in row #4 of Table 1.

Table 1. Resistances (Ω) from Measurement #1

	R$_1$	R$_2$	R$_3$	R$_4$	R$_a$	R$_b$	R$_c$	R$_d$	R$_e$	R$_f$
measured resistance (Ω) using DMM										
formula of equivalent resistance R_{eq}	--	--	--	--						
calculated equivalent resistance R_{eq}	--	--	--	--						

Note: R_{eq} refers to R$_a$, R$_b$, R$_c$, R$_d$, R$_e$, and R$_f$ in Table 1.

Questions on Measurement #1:

Are the calculated values in row #4 in agreement with the measured values in row #2 of Table 1?

Caution: **Turn off the power supply before you make connections. Don't let the two output terminals of the power supply touch each other anytime.**

Measurement #2: use Voltmeter to measure dc voltages across the elements in Fig. 2

1. Check that the DC power supply is off. Connect the circuit shown in Fig. 2.
2. Turn on the power supply and measure the voltage:
   (a) V$_1$ across R$_1$;   (b) V$_2$ across R$_2$;   (c) V$_S$ across the power supply.
3. Record the measured data in Table 2.
4. Turn voltage to 0V then shut down the power supply.

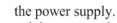

Figure 2

Table 2 Voltages from Measurement #2

V$_S$ (V)	V$_1$ (V)	V$_2$ (V)

Questions on Measurement #2:

In Table 2, is V$_1$ + V$_2$ = V$_S$? Why?

Measurement #3: use Ammeter to measure dc current through the loop in Fig. 3

1. Check that the DC power supply is off. Connect the circuit shown in Fig. 3
2. Turn on the power supply to measure the current through the loop, record it in Table 3.
3. Turn off the power supply.

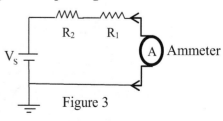

Figure 3

Table 3 Current through the loop from Measurement #3

V$_S$ (V)	I (A)

Questions on Measurement #3:
Can you connect the ammeter between R_1 and R_2 to measure the current through the loop in Fig. 3? What value of the current do you expect to get? Why?

Measurement #4: use Voltmeter to measure dc voltages across the elements in Fig. 4
1. Check that the DC power supply is off. Connect the circuit shown in Fig 4.
2. Turn on the power supply, measure the following voltages:
   (a) V_1 across R_1;   (b) V_2 across R_2;   (c) V_S across the power supply.
3. Record your data in Table 4.
4. Turn off the power supply.

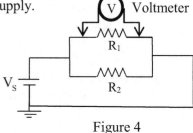

Table 4 Voltages from Measurement #4

V_S (V)	V_1 (V)	V_2 (V)

Questions on Measurement #4:
In Table 4, is $V_1 + V_2 = V_S$?   Why?   Is $V_1 = V_2 = V_S$?   Why?

Measurement #5: use Ammeter to measure dc current through the loop in Fig. 5
1. Check that the DC power supply is off. Connect the circuit shown in Fig. 5.
2. Turn on the power supply, measure the following currents:
   (a) I_1 through R_1,   (b) I_2 through R_2,   (c) I_S through the power supply.
3. Record your data in Table 5.
4. Turn off the power supply.

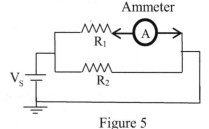

Table 5 Currents from Measurement #5

I_S (A)	I_1 (A)	I_2 (A)

Questions on Measurement #5:
In Table 5, is $I_1 = I_2 = I_S$?   Why?   Is $I_1 + I_2 = I_S$?   Why?

Measurement #6: use voltmeter and ammeter to measure dc voltage and current in Fig. 6
1. Turn off the DC power supply, connect the circuit shown in Fig. 6.

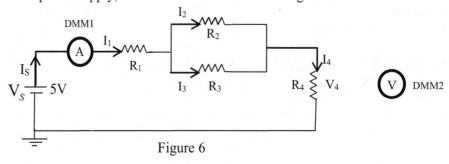

Figure 6

26

2. Turn on the DC power supply, measure its output voltage with DMM2 and record it as measured data V_S in Table 6.
3. Use DMM2 to measure the voltages V_1, V_2, V_3, V_4 across R_1, R_2, R_3, R_4 respectively. Use DMM1 to measure the current I_s (notice: $I_S = I_1 = I_2 + I_3 = I_4$). Record the measured values of I_s and V_1, V_2, V_3, V_4 in Table 6.
4. Turn off the DC power supply.
5. Use relevant formulas (in Theory of Experiment 2) to calculate the voltages V_1, V_2, V_3, V_4 and currents I_S, I_1, I_2, I_3, I_4 using $V_S = 5V$ and the values of R_1, R_2, R_3, R_4 in Table 1. Record all the calculated data in Table 6.

Table 6 Measured and calculated voltages and currents

	Measured data	Calculated data
Source voltage V_S		5V
V_1 (V)		
V_2 (V)		
V_3 (V)		
V_4 (V)		
I_S (mA)		
I_1 (mA)		
I_2 (mA)	Not measured	
I_3 (mA)	Not measured	
I_4 (mA)		

Questions on Measurement #6:
Do the calculated values agree with the measured values in Table 6?

Work to be done:
1. Check that the DC power is off.
2. You may tear those pages containing the raw data Tables out of the lab manual and used them as parts of your lab report.
3. Let your TA check your raw data Tables. If they are OK, your TA will sign them.
4. Clean up your bench.

Lab report on Experiments 3
1. Your lab report should be in the required format described in the "Introduction" of the lab manual.
2. Tables 1 to 6 along with their corresponding circuits should be included in your lab report.
3. It is required that the answers to the questions on each measurement listed below each Table, and in "Questions and Exercises" should be included in your lab report.
4. You can tear those pages out of the lab manual as a part of your lab report, which contain measured (raw) data and analyzed data, answers to questions. The data sheets must be checked and signed by your lab TA.

Questions and Exercises

1. Choose yes or no in the answers to the following questions.
   - (a) To measure the resistance of a resistor in a circuit you need to break the circuit, i.e., pull out a leg of the resistor from the circuit. Answer: Yes (✓), No ( ).
   - (b) To measure the voltage across a resistor in a circuit you need to break the circuit, i.e., pull out a leg of the resistor from the circuit. Answer: Yes ( ), No (✓).
   - (c) To measure the current through a resistor in a circuit you need to break the circuit, i.e., pull out a leg of the resistor from the circuit. Answer: Yes ( ), No (✓).

2. Choose yes or no in the answers to the following questions.
   - (a) To measure the voltage across a resistor in a circuit you need to connect a **Voltmeter** parallel to the resistor. Answer: Yes (✓), No ( ).
   - (b) To measure the current through a resistor in a circuit you need to connect an **Ammeter** parallel to the resistor. Answer: Yes ( ), No (✓).

3. What is wrong in Fig. 7 for measuring the resistance of R_1 using an **Ohmmeter**? Make a correct drawing.

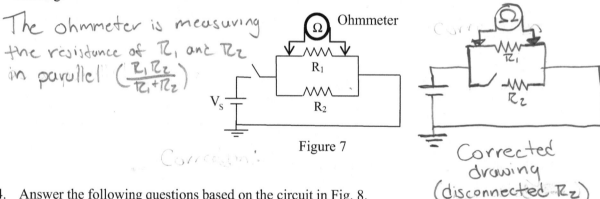

The ohmmeter is measuring the resistance of R_1 and R_2 in parallel $\left(\frac{R_1 R_2}{R_1 + R_2}\right)$

Figure 7

Corrected drawing (disconnected R_2)

4. Answer the following questions based on the circuit in Fig. 8.
   - (a) When switches S_1 and S_2 are open, the ammeter reading is 0.6 A and voltmeter reading is 2.4 V calculate:
     - the resistance of the lamp; $2.4 = 0.6 \cdot R$, $R = 4\Omega$
     - the resistance of R_1. $6 - 2.4 = R_{1V} = 3.6$, $3.6 = 0.6 \cdot R$, $R = 6\Omega$
   - (b) Compare the brightness of the lamp when both switches are open as in (a) and when only switch S_1 is closed. Give a quantitative explanation.
   - (c) Compare the brightness of the lamp when both switches are open as in (a) and when both switches S_1 and S_2 are closed. Give a quantitative explanation.
   - (d) For $R_2 = 4\ \Omega$, if switch S_1 is opened but switch S_2 is closed, what is ammeter reading? What is voltmeter reading?

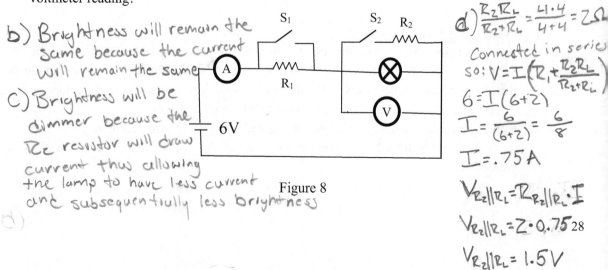

b) Brightness will remain the same because the current will remain the same

c) Brightness will be dimmer because the R_2 resistor will draw current thus allowing the lamp to have less current and subsequentially less brightness

Figure 8

d) $\frac{R_2 R_L}{R_2 + R_L} = \frac{4 \cdot 4}{4 + 4} = 2\Omega$

Connected in series

so: $V = I\left(R_1 + \frac{R_2 R_L}{R_2 + R_L}\right)$

$6 = I(6 + 2)$

$I = \frac{6}{(6+2)} = \frac{6}{8}$

$I = .75A$

$V_{R_2 \| R_L} = R_{R_2 \| R_L} \cdot I$

$V_{R_2 \| R_L} = Z \cdot 0.75$

$V_{R_2 \| R_L} = 1.5V$

28

Experiment 4
Voltmeter and ammeter

Purpose
Construct a voltmeter (0 to 10 V) and an ammeter (0 to 100 mA) using a galvanometer and resistors.

Equipment
Galvanometer (500 μA), resistance box, R_m (= R_1 or R_2 or R_3, multiplier resistors for a lab-made voltmeter), ruler (30 cm), steel wire (shunt resistor for a lab-made Ammeter), digital multmeter (DMM), potentiometer (100 Ω to 50 KΩ), DC power supply (HY152A).

Theory
Galvanometer is the basic indicating component of all deflection-type meters. It is an electromagnetic device capable of detecting very small currents. The basic design of a moving coil galvanometer is shown in Fig. 1.

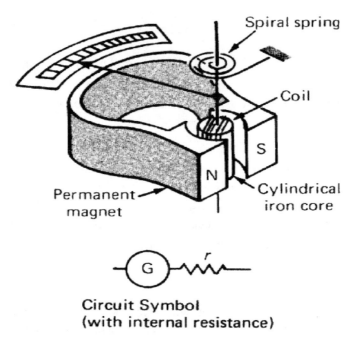

Circuit Symbol
(with internal resistance)

Figure 1 Galvanometer

Galvanometer consists of a coil of wire mounted on bearings between two poles of a permanent magnet. When a current passes through the coil, it experiences a torque and rotates the pointer along the scale until balanced by the torque supplied by two small springs. The deflection is proportional to the current in the coil.

With an additional resistor and a new scale it can become a voltmeter or an ammeter. The galvanometer used for meter construction is characterized by its **internal coil resistance r** and **coil current I_c** required to give a full-scale deflection of the pointer. Once r and I_c are known the additional resistance necessary for converting a galvanometer to a desired voltmeter or ammeter can be calculated.

1. dc Voltmeter

The goal here is to design a circuit that can be connected across a load resistor R that will allow a small current that is proportional to the voltage drop across R to flow through the galvanometer, and that will divert as little current as possible from the load resistor R. The circuit in Fig. 2 accomplishes both these goals.

The series resistor R_m is called a **multiplier resistor** because it multiplies the voltage range of the galvanometer. The combination of a galvanometer with a multiplier resistor makes a voltmeter which is usually put in a single box.

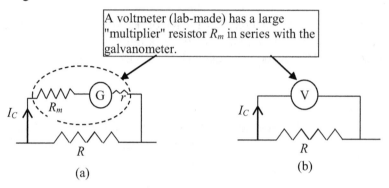

Figure 2   Voltmeter

The multiplier resistance R_m can be calculated using Ohm's law. If you want to construct a voltmeter with a maximum scale reading V_{max}, the total resistance of the voltmeter equals V_{max}/I_C. Since the multiplier resistance R_m and the internal resistance r of the galvanometer are in series, therefore we have:

$$R_m = \frac{V_{max}}{I_C} - r \qquad\qquad (1)$$

Note: A voltmeter is always connected "across" or in parallel with a circuit component such as R in Fig. 2(b) in order to measure the potential difference or voltage drop across the component. Notice that **you don't need to break the circuit** for measuring the voltage across an element in a circuit. If a voltmeter were connected in series with a circuit component, its high resistance (R_m) would reduce the current in the circuit as well as the voltage drop across that component.

2. dc Ammeter

The challenge here is to design a circuit that will divert a small amount of current through the galvanometer that is proportional to much larger current flowing through the load resistor R. This diversion is accomplished by placing a small **"shunt" resistor R_{sh}** in parallel connection with the internal resistance r of the galvanometer as shown in Fig. 3.

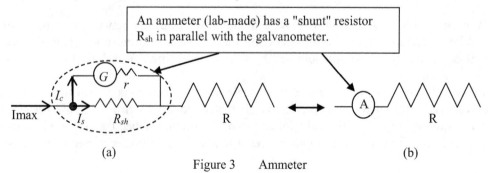

Figure 3   Ammeter

The small resistor connected in parallel with the galvanometer is called a shunt because it directs or "shunts" most of the current away from the galvanometer. As the current flows through the shunt resistance R_{sh}, a small voltage is developed that diverts a small amount of current through the galvanometer, causing it to have a controlled deflection proportional to the current through the circuit. The combination of a galvanometer with a shunt resistor makes an ammeter.

Shunt resistance required for a full-scale deflection of a desired maximum current can be calculated using Ohm's law. With a full scale deflection, the current through the galvanometer is I_c and the voltage drop across the galvanometer is $I_c\, r$. Since the galvanometer and the shunt resistor are in parallel, the voltage across the shunt resistor and the galvanometer should be the same:

$$I_c\, r = I_s\, R_{sh} \qquad\qquad (2)$$

R_{sh} is much smaller than r; therefore, the current through the shunt must be much larger than the current through the galvanometer. If you want to construct an ammeter with a maximum scale reading I_{max} then the maximum current flowing in the load resistor R equals I_{max}. Since $I_s = I_{max} - I_c$, Eq. (2) can be solved for R_{sh}:

$$R_{sh} = \frac{I_c\, r}{I_{max} - I_c} \qquad\qquad (3)$$

Note: An **ammeter is always connected in series with a circuit component** (Fig. 3 (b)) in order to measure the current flowing through that component, thus, **you need to break the circuit**, e.g., pull out a leg of R from the circuit in Fig. 3(b), then insert the ammeter at the break point so that the R and ammeter is in series. If an ammeter were connected in parallel with a circuit component, its low resistance would make the circuit component to be electrically shorted, as a result, a large current would pass through the circuit, the ammeter and the power supply could be damaged.

Procedure

1. Use a DMM to measure the internal resistance r of the Galvanometer, record the measured value in Table 1. The **full-scale current I_c** of the Galvanometer used in this experiment is $500\,\mu A$ which is listed in Table 1.

Table 1 Internal resistance r and full-scale current I_c of the Galvanometer

Galvanometer	r (Ω)	I_c (μA)
	123	500

2. **Build and test a dc Voltmeter**
    (a) Use Eq. (1) to calculate the multiplier resistance (R_m) necessary for constructing a voltmeter with a maximum scale reading $V_{max} = 10$ V.
    (b) Three R_m, i.e., R_1, R_2 and R_3 are to be used in this experiment. Use a DMM to measure the resistance of each R_m and record the data in Table 2.
    (c) Connect R_1 with the galvanometer properly (Fig. 2(a)) to make a 10 V lab-made voltmeter.
    (d) Use both the lab-made voltmeter and a DMM to measure the voltage across the potentiometer in the circuit shown in Fig. 4.

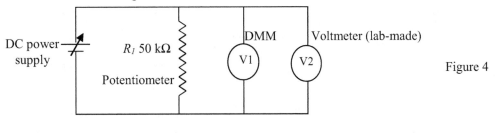

Figure 4

Figure 8

(e) Set the potentiometer to maximum (50 kΩ). Turn on the power supply, start from 0 V, increase the voltage of the power supply so that the readings on V2 (lab-made, $R_m = R_1$) are 2.0, 4.0, ...10.0 V, in the meantime record the corresponding readings from the V1 (DMM) in Table 2.

(f) Turn the power supply to 0 V then shut it down.

(g) Repeat steps (c) to (f) for $R_m = R_2$ and $R_m = R_3$ respectively.

Table 2 Voltage readings from both the lab-made voltmeter and the DMM.

V2 (lab-made, $R_m = R_1$)	2.0 V	4.0 V	6.0 V	8.0 V	10.0 V
V1 (DMM)	2.07	3.81	5.76	7.68	9.67
V2 (lab-made, $R_m = R_2$)	2.0 V	4.0 V	6.0 V	8.0 V	10.0 V
V1 (DMM)	1.53	3.12	4.69	6.14	7.82

V2 (lab-made, $R_m = R_3$)	2.0 V	4.0 V	6.0 V	8.0 V	10.0 V
V1 (DMM)	3.02	5.75	8.44	11.28	14.22

(h) In Table 2, which set of data from the lab-made voltmeter is the best match with the data from the DMM? Why?

(g) What is the voltage across the galvanometer when the reading on V2 (lab-made) is 10.0 V?

3 Build and test a dc Ammeter

(a) Use Eq. (3) to calculate the shunt resistance (R_{sh}) necessary for constructing an ammeter with a maximum scale reading I_{max} = 100 mA.

(b) A small shunt resistor used for building an ammeter is hard to be found from commercially available resistors. A piece of steel wire (0.035 Ω/ cm) is used to make a shunt resistor in this experiment. Calculate the length of the wire so that the wire resistance equals R_{sh}. Record the data in Table 3.

(c) If the calculated length of the wire is ℓ (cm), then you need to cut a piece steel wire with a length of $\ell + 5$ (cm) in order to leave some room for further adjustment (see step (h)) and for connections at the two posts of the galvanometer.

(d) Connect the shunt wire of length $\ell + 5$ (cm) properly at the two posts of the galvanometer (Fig. 3(a)) to make an ammeter (lab-made) with a maximum scale reading I_{max} = 100 mA.

(e) Use both the lab-made ammeter and a DMM to measure the current through the potentiometer in the circuit shown in Fig. 5.

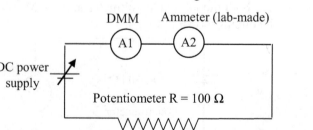

Figure 5

(f) Set the potentiometer to minimum (100 Ω). Turn on the power supply starting from 0 V, increase the voltage of the power supply so that the readings on A2 (lab-made, use a shunt wire of length

32

$\ell + 5 =$ _____ (cm)) are 20.0, 40.0, ...100.0 mA, in the meantime record the corresponding readings from A1 (DMM) in Table 3.

(g) Turn the power supply to 0 V before you turn it off.

(h) Repeat steps (d) to (g) for a shunt wire of length ℓ (cm) which has a calculated resistance R_{sh}.

(i) Repeat steps (d) to (g) for a shunt wire with length of $\ell - 5$ (cm).

Table 3 Current readings from both the lab-made ammeter and the DMM.

A2 (lab-made, use a shunt wire of length $\ell + 5 =$ 22 (cm)	20.0 mA	40.0 mA	60.0 mA	80.0 mA	100.0 mA
A1 (DMM)	15.93	32.04	48.05	63.42	78.32

A2 (lab-made, use a shunt wire of length $\ell =$ 17 (cm)	20.0 mA	40.0 mA	60.0 mA	80.0 mA	100.0 mA
A1 (DMM)	18.18	40.20	57.11	78.72	101.05

A2 (lab-made, use a shunt wire of length $\ell - 5 =$ 12 (cm)	20.0 mA	40.0 mA	60.0 mA	80.0 mA	100.0 mA
A1 (DMM)	23.40	46.92	72.07	94.32	123.00

(j) In Table 3, which set of data from the lab-made ammeter is the best match with the data from the DMM? Why?

(k) What is the current through the galvanometer when the reading on A2 (lab-made) is 100.0 mA?

Work to be done:
1. Check that the DC power (HY152A) is off. If not, turn voltage to 0V then shut it down.
2. Let your TA check your data Tables. If they are OK, your TA will sign them.
3. Clean up your bench.

Lab report on Experiments 4
1. Your lab report should be in the required format described in the "Introduction" of the lab manual.
2. Tables 1 to 3 and the circuits used to test the lab-made Voltmeter and Ammeter should be included in your lab report.
3. It is required that the answers to the questions listed below Tables 2, 3 and in "Questions and Exercises" should be included in your lab report.
4. You can tear those pages out of the lab manual as a part of your lab report, which contain measured (raw) data and analyzed data, answers to questions. The data sheets must be checked and signed by your lab TA.

Questions and exercises

1. This is a real story. A student tried to test her lab-made Ammeter with a DMM using the circuit in Fig. 6. As soon as she closed the switch the power supply started burning and her finger was injured.

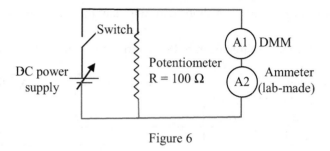

Figure 6

(a) What is wrong with the circuit in Fig. 6?

(b) Draw a correct circuit for testing a lab-made Ammeter with a DMM.

2. Suppose you have a galvanometer with a full-scale current $I_c = 50$ μA and an internal resistance $r = 200$ Ω.
   (a) What resistance value of a resistor should be used for making a dc voltmeter with a maximum scale reading $V_{max} = 20$ V?

   (b) What resistance value of a resistor should be used for making a dc ammeter with a maximum scale reading $I_{max} = 200$ mA?

Experiment 5
RC time constant

Purpose
1. Investigate voltage variations in the processes of charging and discharging a capacitor.
2. Study time constant of a RC circuit.

Equipment
Two resistors $R_1 \cong R_2 \cong 10\,k\Omega$, two capacitors $C_1 \cong C_2 \cong 1\,\mu C$, one digital multimeter (DMM), one voltage sensor (PASCO UI-5100), PASCO 850 interface, computer. Demo box of capacitors.

Theory
When a DC voltage source is connected across an uncharged capacitor (switch in position a in Fig. 1a), the rate at which the capacitor charges up decreases as time passes. At first, the capacitor is easy to charge because there is very little charge on the plates. But as charge accumulates on the plates, the battery (or dc power supply) must "do more work" to move additional charges onto the plates because the plates already have charges of the same sign on them. As a result, the capacitor charges exponentially, quickly at the beginning and more slowly as the capacitor becomes fully charged. The charge on the plates at any time is given by:

$$q = q_0(1-e^{-t/RC}) = q_0(1-e^{-t/\tau}),$$

where q_0 is the maximum charge on the plates and $\tau = RC$ is the capacitive time constant (R is resistance and C is capacitance).

In this experiment the charge on the capacitor is measured indirectly by measuring the voltage across the capacitor since these two values are proportional to each other: $q = CV$.

At time $t = \tau = RC$, the voltage across the capacitor has grown to a value (as shown in Fig.1b):

$$V = V_0(1-e^{-1}) = 0.63\,V_0 ,$$

where V_0 is the maximum voltage across the capacitor.

When fully charged capacitor is discharged through a resistor (switch in position b in Fig. 1a), the voltage across the capacitor "decays" or decreases with time according to the equation:

$$V = V_0\,e^{\frac{-t}{RC}} = V_0\,e^{\frac{-t}{\tau}} .$$

After a time $t = \tau = RC$, the voltage across the capacitor has decreased to a value (as shown in Fig. 1b):

$$V = V_0\,e^{-1} = \frac{V_0}{e} = 0.37\,V_0.$$

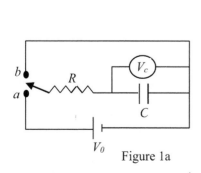

Figure 1a

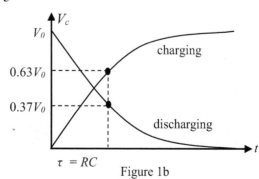

Figure 1b

Figure 1 RC circuit: process of charging and discharging a capacitor

Measurements to be done

Notation: R_p (R_s) is the equivalent resistance of R_1 and R_2 in parallel (series) connection,

C_p (C_s) is the equivalent capacitance of C_1 and C_2 in parallel (series) connection.

Six measurements are to be done using computer.

Measurement #1: Time constant τ of R_pC_1 circuit using a square wave at $f = 5$ Hz with $V_0 = 4$ V.

Measurement #2: Time constant τ of R_1C_1 circuit using a square wave at $f = 5$ Hz with $V_0 = 4$ V.

Measurement #3: Time constant τ of R_sC_1 circuit using a square wave at $f = 5$ Hz with $V_0 = 4$ V.

Measurement #4: Time constant τ of R_1C_p circuit using a square wave at $f = 5$ Hz with $V_0 = 4$ V.

Measurement #5: Time constant τ of R_1C_s circuit using a square wave at $f = 5$ Hz with $V_0 = 4$ V.

Measurement #6: Charging and discharging processes of R_1C_1 circuit using a square wave at $f = 50$ Hz with $V_0 = 4$ V.

Preparation

1. Measure resistance
    (a) Measure the resistances of R_1 and R_2 using DMM, record the measured values in Table 1.
    (b) Connect R_1 and R_2 in parallel, measure the equivalent resistance R_p and record it in Table 1. R_p will be used in Measurement #1.
    (c) Connect R_1 and R_2 in series, measure the equivalent resistance R_s and record it in Table 1. R_s will be used in Measurement #3.
2. Calculate the equivalent capacitance
    (a) Calculate the equivalent capacitance C_p of C_1 and C_2 in parallel connection using formula

    $C_p = C_1 + C_2$ and record it in Table 1. C_p will be used in Measurement #4.

    (b) Calculate the equivalent capacitance C_s of C_1 and C_2 in series connection using formula

    $\dfrac{1}{C_s} = \dfrac{1}{C_1} + \dfrac{1}{C_2}$ and record it in Table 1. C_s will be used in Measurement #5.

Table 1 Measured values of resistances and capacitances

R_1	R_2	R_s	R_p	C_1 (μF)	C_2 (μF)	C_s (μF)	C_p (μF)

3. Connect the RC circuit as shown in Fig. 2.
    The sketch of PASCO 850 interface is in the "Introduction" of this lab manual.

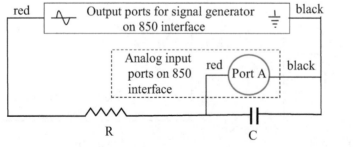

Figure 2
RC circuit diagram

Caution: **Turn off the power of PASCO 850 interface before you make connections. Don't let the two output terminals from the output ports on 850 interface touch each other anytime.**

(a) Use $R = R_p$ and $C = C_1$ to connect the circuit in Fig. 2 for measurement #1.

(b) Use one red and one black banana leads to connect the RC series and the Output Ports (banana jacks) of the signal generator on 850 Interface as shown in Fig. 2.

(c) Use a voltage sensor to connect the capacitor and Port A of the Analog Inputs on 850 Interface.

Attention: the two black leads must be connected to the C-end of the RC series as shown in Fig. 2.

4. Turn on PASCO 850 Interface and computer. Log in computer using Password: phy2049

5. Start PASCO Capstone by double clicking the symbol on computer.
The Workbook Page appears.

6. Hardware Setup on the Workbook Page

(a) In the Tools Palette, click on the "Hardware Setup" icon, the Hardware Setup panel appears with a picture of the PASCO 850 Interface along with a 'voltage sensor' icon [⟋] on the top of Port A on the Analog Inputs.

Attention: If the 'voltage sensor' icon does not show up automatically, click Port A on the Analog Inputs (in the PASCO 850 Interface picture), a drop down menu of sensors appears. Type in "v", in the pop-up select "voltage sensor", then a 'voltage sensor' icon appears.

(b) Click the Output Ports on the upper-right corner of the picture of 850 Interface, in the pop-up select the "Output Voltage-Current Sensor", the sensor icon appears on the top.

(c) In the Controls Palette (at the bottom of the Workbook Page)
- select "Continuous Mode",
- set sampling rate to 20 kHz,
- click "Recording conditions", select "Stop condition"/"Time base" and set to 0.4 s.

(d) Click on "Hardware Setup" icon to close the Hardware Setup panel.

7. Setup Signal Generator on the Workbook Page

(a) In the Tools Palette, click "Signal Generator" button to open the Signal Generator panel.

(b) Click on "850 Output 1" and make the following selections:
- Wave form: Square
- Sweep: Off
- Frequency: 5 Hz
- Amplitude: 2 V
- Voltage Offset: 2 V
- Auto

(c) Click on "Signal Generator" icon to close the Signal Generator panel.

Note: A square wave with 2 V amplitude and 2 V offset is equivalent to a square wave all positive with an amplitude $V_0 = 4$ V.

8. Setup graph display Page #1 on the Workbook page:

(a) Double click the graph icon in the Displays palette to open graph Page #1 on Workbook page.
Note: If the left side of the graph Page #1 is blocked by any Tools panel such as "Hardware Setup panel" or Signal generator panel", click on the red pin ⚲ at the upper-right corner of the Tools panel, then the full graph Page #1 is displayed on the right side of the Tools panel.

(b) Click on the y-axis label and in the pop-up select "Output Voltage, Ch 01 (V)" --- the voltage of the square wave $v_s(t)$ from the Signal generator which is applied to the RC series.

(c) Click "Add new y-axis to active plot area" on the top of the graph Tool bar, the new y-axis is displayed on the right side of graph Page #1. Set the new y-axis to Ch A (V) --- the voltage $v_C(t)$ across the capacitor C.

Measurement #1: Time constant τ of R_pC_1 circuit using a square wave at $f = 5$ Hz with $V_0 = 4$ V.
1. Click "Record" button in the Controls palette on the lower-left corner of the Workbook Page to start data collection which will be automatically stopped at 0.4 s. Two cycles of $v_s(t)$ and $v_C(t)$ curves are displayed on graph Page #1.
2. Click on the "Data Summary" button at the left edge of the page, right click on Run #1 and rename it as "R_pC_1-5Hz".
3. Turn off 850 interface.

Measurement #2: Time constant τ of R_1C_1 circuit using a square wave at $f = 5$ Hz with $V_0 = 4$ V.
1. Use $R = R_1$ and $C = C_1$ to connect the circuit in Fig. 2.
2. Repeat steps in Measurement #1 to perform Measurement #2. Rename Run #N as "R_1C_1-5Hz".

Measurement #3: Time constant τ of R_sC_1 circuit using a square wave at $f = 5$ Hz with $V_0 = 4$ V.
1. Use $R = R_s$ and $C = C_1$ to connect the circuit in Fig. 2.
2. Repeat steps in Measurement #1 to perform Measurement #3. Rename Run #N as "R_sC_1-5Hz".

Measurement #4: Time constant τ of R_1C_p circuit using a square wave at $f = 5$ Hz with $V_0 = 4$ V.
1. Use $R = R_1$ and $C = C_p$ to connect the circuit in Fig. 2.
2. Repeat steps in Measurement #1 to perform Measurement #4. Rename Run #N as "R_1C_p-5Hz".

Measurement #5: Time constant τ of R_1C_s circuit using a square wave t $f = 5$ Hz with $V_0 = 4$ V.
1. Use $R = R_1$ and $C = C_s$ to connect the circuit in Fig. 2.
2. Repeat steps in Measurement #1 to perform Measurement #5. Rename Run #N as "R_1C_s-5Hz".

Measurement #6: Charging and discharging processes of R_1C_1 circuit using a square wave at $f = 50$ Hz with $V_0 = 4$ V.
1. Use $R = R_1$ and $C = C_1$ to connect the circuit in Fig. 2.
2. In Signal Generator panel, set Frequency: 50 Hz.
3. In "Recording conditions", set "Stop condition"/"Time base" to 0.1 s.
4. Click "Record" button to start data collection which will be automatically stopped at 0.1 s. Four cycles of $v_s(t)$ and $v_C(t)$ curves are displayed on graph Page #1.
5. Rename Run #N as "R_1C_1-50Hz".

Simultaneous viewing of multiple data sets on graph Page #1 and Page #2 respectively

Now total 6 sets of data ($v_s(t)$ and $v_C(t)$ curves) have been collected from above 6 measurements.
1. Display simultaneously data sets #1 to #5 (from Measurements #1 to #5) on Page #1
   Click the Data Selection Tool (▲▼) on the graph tool bar so that it is pushed in, then click on the small black triangle and check the names of the 5 sets of data.
   If you don't want to display some of the sets of data, just uncheck their names.
   The display should be similar to Fig. 3.
2. Open graph Page #2 by clicking "Add Page" on the left of Page #1. Setup Page #2 in the same way as for Page #1 in **Preparation** (step 8).
3. Use above procedure to display data sets #2 and #6 on Page #2 which should be similar to Fig. 4.

Data Analysis 1: Analyze data on graph Page #1

1. Click on any spot on graph Page #1 and drag it to left. Leave a space between the left side of the square wave and the y-axis.
2. Change the x-axis scale by moving the hand cursor above a number on the x-axis scale, when the hand cursor changes to horizontal cursor ⇔, click and drag the ⇔ cursor to right until only one complete group is displayed on Page #1 similar to Fig. 3.
3. Similarly, change the y-axis scale by dragging the vertical cursor ↕ downward along the y-axis to make a space between the top line of the square wave and the top frame of Page #1.
4. Click on any spot on the graph and drag it upward to make a space between the bottom line of the square wave and the x-axis.

    Now your graph Page #1 should be similar to Fig. 3 which shows the first complete group including one source square wave $v_s(t)$ and five $v_C(t)$ curves.

5. There is a legend (an information table) on the upper right corner of the graph Page #1. You can click on the frame of the legend and drag it to a proper place. The first column of the legend lists the names of the five $v_C(t)$ curves defined in **Measurements #1 to #5**. The second column lists the source voltage $V_0 = V_s(t)$ (the square wave). The third column lists the voltage across the capacitor $V = V_C(t)$.

6. Label coordinates on the five $v_C(t)$ curves

    (a) Two points on each $v_C(t)$ curve to be labeled:

    - Point 1 at $v_C(t_1) = 0V$ --- the beginning of the charging process, no charge on C yet.

        Note: The five $v_C(t)$ curves have a common Point 1 whose coordinates can be read directly from the display on Page #1. In Fig. 3, the coordinates of Point 1 is $[t_1 = 0.30\ s,\ v_C(t_1) = 0V]$. Record t_1 in Table 2.

    - Point 2 is at $v_C(t_2) = 0.63V_0 = 0.63 \times 4 = 2.52V$ at which the capacitor is charged up to 63% of the maximum value (V_0). The info-box of Point 2 is $[t_2 = \quad s,\ v_C(t_2) = 2.52V]$.

    (b) Use the Coordinates Tool (⊹) to label Point 2 on each $v_C(t)$ curve

    Click on one of the five $v_C(t)$ curves to highlight it (the curve becomes active), then click on the Coordinates Tool (called Smart Cursor) (⊹) from the graph Toolbar, an empty dashed square box and an info-box with two values [time, voltage] appear. Click on the empty square box and drag it along the highlighted $v_C(t)$ curve until the voltage value in the info-box is equal or very close to 2.52 V, then record the time value of t_2 (read from the info-box) in Table 2.

7. $\tau_{exp} = t_2 - t_1$ is the experimental RC time constant. Record τ_{exp} in Table 2.

8. Use $\tau_{thoery} = RC$ to calculate the theoretical RC time constant and record τ_{theory} in Table 2.

Table 2 Experimental and theoretical RC time constants using a square wave with $f = 5\,Hz$, $V_0 = 4V$

Measurement	f (Hz)	t_1 (s)	t_2 (s)	τ_{exp} (s)	R (kΩ)	C (µf)	τ_{theory} (s)
#1 R_pC_1-5Hz	5						
#2 R_1C_1-5Hz	5						
#3 R_sC_1-5Hz	5						
#4 R_1C_p-5Hz	5						
#5 R_1C_s-5Hz	5						

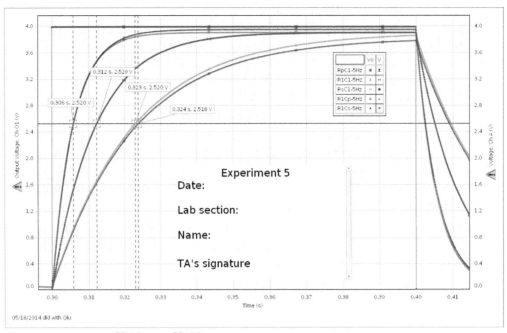

Figure 3 $V_s(t)$ and $V_C(t)$ curves obtained from Measurements #1 to #5

Data Analysis 2: Analyze data on graph Page #2

1. Use the Coordinates Tool ( ) to label the first two charging processes on the R_1C_1-50Hz $\upsilon_C(t)$ curve on the left side of graph Page #2. The following notation of the info-box should be used: $\left[t_{min}, \upsilon_C(t_{min})\right]$ and $\left[t_{max}, \upsilon_C(t_{max})\right]$ where "min" is at the starting point of a charging process and max is at the maximum υ_C value of the charging process. Notice that the $\left[t_{min}, \upsilon_C(t_{min})\right]$ of the first charging process can be read directly from the R_1C_1-50Hz $\upsilon_C(t)$ curve, so that only three points need to be labeled as shown in Fig. 4. Record the labeled values in Table 3.

2. Repeat step 1 to label the R_1C_1-5Hz $\upsilon_C(t)$ curve.

3. Your graph Page #2 should be similar to Fig. 4.

Table 3 Data from Measurements # 2 and #6 using R$_1$ = _____ kΩ, C$_1$ =_____μf, V$_0$ = 4 V, f = 5 Hz and 50 Hz respectively.

Measurement #2	V_{max} (V)	V_{min} (V)	$V_{max} - V_{min}$ (V)	Is C_1 fully charged?
R_1C_1-5Hz				

Measurement #6	V_{max} (V)	V_{min} (V)	$V_{max} - V_{min}$ (V)	Is C_1 fully charged?
R_1C_1-50Hz				

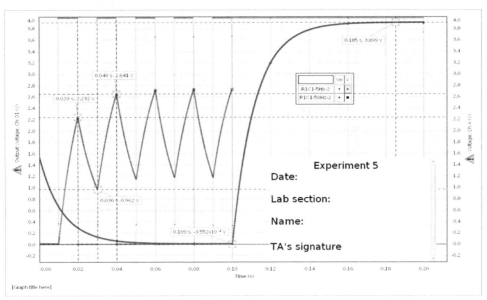

Figure 4 $V_s(t)$ and $V_C(t)$ curves obtained from Measurements #1 and #6

4. Create a Text Box on graph Page #1 (#2)
   (a) Double click the Text Box icon in the Displays palette to open it on Page #1 (#2).
   (b) In the Text Box type in the contents as shown in Figs. 3 and 4.
       When you put mouth cursor at any point in the Text Box a command ribbon appears at the top of Text Box. You can change the font and font size by click "A" and in the pop-up select your favor font and font size.
   (c) You may reduce the size of the Text Box by dragging an arrow cursor (⇔) along the diagonal from a corner of the blue frame.
   (d) You can move the Text Box by clicking on and dragging one of the four sides (blue frame) to where you want to place it.

Work to be done:
1. Print Graph Page #1 (similar to Fig. 3) and Graph Page #2 (similar to Fig. 4) as a part of your lab report on Experiment 5.
   Click on "File" at the top-left" corner, in the pop-up select "Print Page Setup" and "Landscape", then print it out.
2. Let your TA check your Pages #1 and #2, if they are OK your TA will sign them.
   Caution: Make sure that your printed Pages #1 and #2 are OK before you close the Capstone. All the data collected in Experiment 5 will be lost when the Capstone is closed!
3. Close PASCO Capstone by clicking the red-cross at the upper corner, then select "discard".
4. Turn off PASCO 850 interface and shut down the computer.
5. Clean up your bench.

Lab report on Experiments 5

1. Your lab report should be in the required format described in the "Introduction" of the lab manual.
2. Graph Page #1 (similar to Figs. 3) and Graph Page #2 (similar to Fig. 4) should be included in your lab report.
3. Tables 1, 2 and 3 should be included in your lab report.
4. It is required that the answers or solutions to the questions (see below) should be included in your lab report.
5. You can tear those pages out of the lab manual as a part of your lab report, which contain measured (raw) data and analyzed data, answers to questions. The data sheets must be checked and signed by your lab TA.

Questions

1. What conclusion on the time constant can be drawn from comparison of **Measurement 2** with **Measurement 4**?

2. What conclusion on the time constant can be drawn from comparison of **Measurement 2** with **Measurement 5**?

3. The circuit contains an ideal battery, two resistors and a capacitor ($C = 250$ μF). The switch is closed at time $t = 0$, and the voltage across the capacitor is recorded as a function of time as shown in the graph.

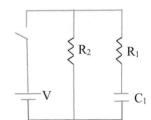

Read the data in the graph carefully and answer the following questions.

(a) What is the voltage of the battery? $V =$ _____.

(b) What is the time constant for this circuit when the switch is closed? $\tau_{closed} =$ _____.

(c) What is the time constant for this circuit when the switch is open? $\tau_{open} =$ _____.

(d) What is the resistance of the resistor R_1? $R_1 =$ _____.

(e) What is the resistance of the resistor R_2? $R_2 =$ _____.

(f) What is the voltage across resistor R_1 at $t = 2.0$ seconds? $V_1 =$ _____.

(g) What is the voltage across resistor R_2 at $t = 2.0$ seconds? $V_2 =$ _____.

(h) What is the total current produced by the battery at $t = 0$? $I =$ _____.

(i) What is the current through resistor R_1 at $t = 5.0$ seconds ? $I_1 =$ _____.

(j) What is the charge on the capacitor $t = 15.0$ seconds ? $q =$ _____.

(k) When is the switch opened? $t =$ _____.

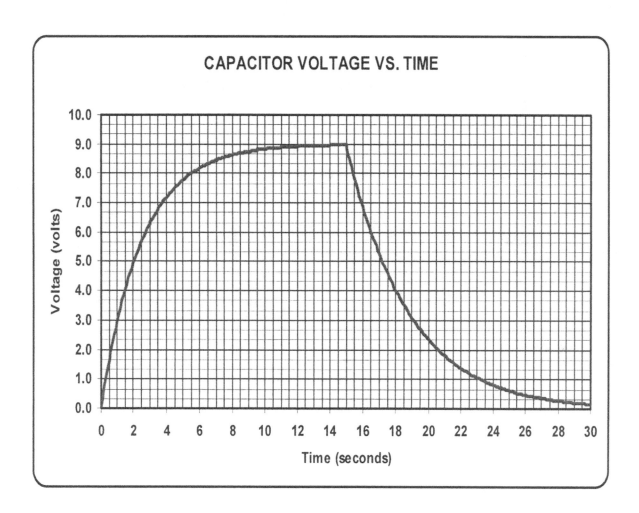

Experiment 6.1
Earth's magnetic field

Purpose

Obtain the horizontal component of the earth's magnetic field (B_h) by measuring the period of oscillation of a bar magnet in a known field.

Equipment

PASCO Helmholtz coils (EM-6711and EM-6715 N=200, radius R=10.5 cm), bar magnet, photogate (ME-9204A), resistor ~ 40 Ω (2W), PASCO 850 interface, computer.

Theory

Except at the magnetic equator, the earth's magnetic field is not horizontal. In South Florida, the magnitude of the earth's field is about 0.5 Gauss while horizontal component is about 0.25 gauss.

The horizontal component of the earth's magnetic field (B_h) can be obtained by studying how it affects a bar magnet which is suspended so that it is free to rotate in a horizontal plane. Such a magnet will line up in the direction of the horizontal component of the earth's field. That is what a compass is!

The reason for this is that a magnetic field of magnitude B will produce a torque τ on a bar magnet given by:

$$\tau = -m\,B\,sin\theta, \qquad (1)$$

where the minus sign means that it is a restoring torque, m is the magnitude of the magnetic moment of the bar magnet, θ is the angle between the magnetic moment and magnetic field vector, as shown in Fig. 1.

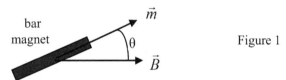

bar magnet $\vec{m}$ θ $\vec{B}$ Figure 1

If the angle θ is small, then $sin\theta \approx \theta$ and equation (1) reduces to

$$\tau = -m\,B\,\theta \qquad (2)$$

This form is similar to $F = -k\,x$ in the case of a simple harmonic oscillator such as a mass hanging from a spring or a simple pendulum.

If we set the bar magnet into oscillation, the motion of the bar magnet is approximately harmonic and the equation of motion is given by the Newton's second law for a rigid body

$$\tau = I\,\alpha = I\frac{d^2\theta}{dt^2}, \qquad (3)$$

where I is the moment of inertia and $\alpha = \dfrac{d^2\theta}{dt^2}$, the angular acceleration of the bar magnet.

Substituting (2) into (3):

$$-m\,B\,\theta = I\frac{d^2\theta}{dt^2}, \qquad \text{or} \qquad \frac{d^2\theta}{dt^2} + \frac{mB}{I}\theta = 0, \qquad \text{or}$$

$$\frac{d^2\theta}{dt^2} + \omega^2\theta = 0 \qquad \text{with} \qquad \omega^2 = \frac{mB}{I}, \qquad (4)$$

where ω is the angular frequency of the oscillation.

45

$$\because \quad \omega = \frac{2\pi}{T},$$

$$\therefore \quad \frac{1}{T^2} = \frac{m}{4\pi^2 I}B = CB, \qquad \text{with} \quad C = \frac{m}{4\pi^2 I}, \qquad (5)$$

T is the period of the oscillation. The constant C depends on the magnetic moment and the moment of inertia of the bar magnet, which is unknown. Hence, in order to obtain the horizontal component of the earth's field, we must first find C for the bar magnet, i.e., we must calibrate it in a known uniform magnetic field.

In this experiment the known uniform magnetic field is produced by Helmholtz coils consisting two identical coils separated by a distance equal to their radius. The two coils carry the same current in a direction such that, at the midpoint between them on their axis, their magnetic field will have the same direction and the same magnitude, so the magnitude of the total magnetic field at that point is

$$B = \left(\frac{4}{5}\right)^{\frac{3}{2}} \frac{N\mu_0 I_C}{R} \approx \frac{0.72 N\mu_0 I_C}{R}, \qquad (6)$$

N is the total # of turns of each Helmholtz coil, R is the radius of Helmholtz coil, I_C is the DC current going through each Helmholtz coil, $\mu_0 = 4\pi \times 10^{-7}$ T·m/A. The SI units (international system of units) of the magnetic units is T (Tesla). The Gaussian units of the magnetic field is G (Gauss). **1 T = 10000 G.**

If we align the Helmholtz coils so that its axis is in the same direction as B_h, the horizontal component of the earth's magnetic field, the fields add to give a total magnetic field $B_{tot} = B_h + \dfrac{0.72 N\mu_0 I_C}{R}$

And equation (5) becomes

$$\frac{1}{T^2} = CB_{tot} = C\left(B_h + \frac{0.72 N\mu_0 I_C}{R}\right) = CB_h + C\frac{0.72 N\mu_0}{R}I_C \qquad (7)$$

This equation shows how we can find the constant C and determine the magnetic field B_h. We cannot obtain C from a single measurement because both C and B_h are unknown. We must measure the period of oscillation T for various values of the current I through the coils. The plot of $\dfrac{1}{T^2}$ versus I_C will be a straight line with a slope $C\dfrac{0.72 N\mu_0}{R}$ and an intercept CB_h from which the values of C and B_h can be determined since the Helmholtz coils used in this experiment with N=200 turns, R =10.5 cm.

Apparatus setup

The sketch of PASCO 850 interface is given in the "Introduction" of this lab manual.
1. To reduce the experimental error the Helmholtz coils should be placed 30 cm away from the computer and the PASCO 850 interface.
2. Place the bar magnet at the center of the Helmholtz coils. When the bar magnet is at rest, its orientation is along the direction of earth's magnetic field. Slightly rotate the Helmholtz coils so that the bar magnet is along the axis of the coils.
3. There is a mark "N" on the base of the Helmholtz coils which indicates the direction of the magnetic field produced by the Helmholtz coils when current flowing through the coils.
   Attention: The mark "N" side of the base should be placed toward north.

Caution: Turn off the power of PASCO 850 interface before you make connections. Don't let the two output terminals from the output ports on 850 interface touch each other anytime.

4. Make connections between the Helmholtz coils and the PASCO 850 interface as shown in Fig. 2.
   (a) Use a red banana lead to connect the middle white banana jack on the right Helmholtz coil (see Fig. 2) and the red banana jack on the Output Ports of PASCO 850 Interface.
   (b) Use a black banana lead to connect the black banana jack on the left Helmholtz coil (see Fig. 2) and the black banana jack on the Output Ports of PASCO 850 Interface.
   (c) Use a red banana lead to connect the front white banana jack on the right Helmholtz coil and the back white banana jack on the left Helmholtz coil (see Fig. 2).

   Attention:
   - The connections shown in Fig. 2 ensure that the direction of the magnetic field produced by the Helmholtz coils is to the mark "N" side. Don't exchange the two red color banana leads on the mark "N" side of the base in Fig. 2.
   - A built-in current sensor of the 850 Interface is used to measure the current flowing through the Helmholtz coils.
5. Connect the phone plug of the photogate to Port 1 of the Digital Inputs on PASCO 850 Interface.

Figure 2 Connections between Helmholtz coils and PASCO 850 Interface

Measurements to be done

Measurement #1: Measure $1/T^2$ vs I_C with mark "N" (on the Helmholtz coils base) pointing north.
   Rename Run #N (N is probably 1) as "North-1"

Measurement #2: Repeat Run #1, rename Run #N as "North-2"

Measurement #3: Repeat Run #1, rename Run #N as "North-3"

Measurement #4: Observation of $1/T^2$ vs I_C with mark "N" (on the Helmholtz coils base) pointing south.

Software setup

1. Turn on PASCO 850 Interface and computer. Log in computer using Password: phy2049
2. Start PASCO Capstone by double clicking the symbol on computer.
   The Workbook Page appears.
3. Hardware Setup on the Workbook Page
   (a) In the Tools Palette, click on the "Hardware Setup" icon, the Hardware Setup panel appears with a picture of the PASCO 850 interface.

(b) Click the Output Ports on the upper-right corner of the picture of 850 Interface, in the pop-up select the "Output Voltage-Current Sensor", the sensor icon ⌷ appears on the top.

(c) Click the Digital Input Port 1 on the upper-left of the picture of 850 Interface, a drop down menu of sensors appears. Type in "ph", in the pop-up select "photogate sensor".

(d) Click on the "Hardware Setup" icon to close the Hardware Setup panel.

4. Timer Setup on the Workbook Page

In the Tools Palette, click on the "Timer Setup" icon to open the Timer Setup panel on which make the following selections. Note: after a selection is made, click "Next" at the bottom of the panel to show the next setting option.

(a) Pre-configured timer

(b) Photogate Ch 1

(c) Pendulum Timer and check the box of "Period"

(d) Pendulum W(0.016m)

(e) Pendulum Timer

Click "Finish", then click "Timer Setup" icon in the Tools Palette to close the Timer Setup panel.

5. Setup Table & Graph Page #1 on the Workbook page:

(a) Click on the Table & Graph icon on the Workbook page, a Table and a Graph display window appear on Graph Page #1.

(b) Setup the Table on Page #1

- In the 1st column of the Table, click "Select measurement", in the pop-up select "Output Voltage (V)".

- In the 2nd column of the Table, click "Select measurement", in the pop-up select "Output Current (A)".

- Click the "Insert empty column" icon at the top-left corner of the Table to add a 3rd column, and click "Select measurement", in the pop-up select "Period (s)".

- Click the "Create a new calculation" icon (#3 icon on the Tools bar of the Table), a 4th column appears. Rename "Calc 1" as "1/T^2".

- In the Tools Palette, click the Calculator icon to open the Calculator panel on which change the existing formula 1/T^2=[Period(s)] to 1/T^2=1/[Period(s), ▼]^2 with units 1/s^2, then click "Acc" at the top of the Calculator panel. The mathematical equation appears below the formula with a message "Measurement assignment OK". The Run # and the title 1/T^2 with units 1/s^2 appear automatically in the 4th column of the Table.

- Click on the Calculator icon to close it.

(c) Set up the Graph display on Page #1

- Click on the y-axis label "Select measurement", in the pop-up select "1/T^2 (1/s^2)".

- Click on the x-axis label "Select measurement", in the pop-up select "Output Current (A)".

Measurement #1: Measure $1/T^2$ vs I_C with mark "N" (on the Helmholtz coils base) pointing north.

1. In the Controls Palette (at the bottom of the Workbook Page) select "Keep Mode".

2. In the Controls Palette set sampling rate for Output Voltage-Current Sensor to 1.0 kHz.

3. In the Tools Palette, click "Signal Generator" button to open the Signal Generator panel.
Click on "850 Output 1" and make the following selections:

(a) Wave form: DC

(b) DC Volt: 0 V

(c) Off

4. The left side of Page #1 is blocked by the "Signal generator panel". Click the red pin ⚲ at the upper-right corner of the Tools panel, then the full Page #1 is displayed on the right side of the "Signal generator panel".

5. Fine adjustment of the bar magnet relative to the photogate
   (a) Adjust the position of the photogate to let one end of the magnet crossing the dashed vertical line connecting the light and receiver of the photogate as shown in Fig. 3, so that the photogate indicator (light) is on and off when the bar magnet passes through the photogate.
   (b) Make sure the bar magnet oscillates **horizontally (**about 30° - 45°).

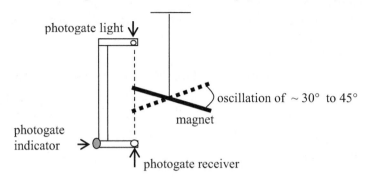

Figure 3 Fine adjustment of the bar magnet relative to the Photogate

6. On Signal Generator panel check DC Volt: 0 V, then click "On" to turn on the DC power supply.
7. Click "Preview" (at the bottom-left of Page #1), the data of voltage, current and period appear in the Table.
8. The oscillation of the bar magnet and the data shown in the Table become stable after the bar magnet makes 2 ~ 3 complete oscillation cycles. Click "Keep Sample" (at the bottom-left of Page #1) to store the data.
9. Increase the DC voltage 1 V each time and repeat steps 5 and 8. The final output voltage from the DC power supply is 10 V, so total 11 data points (from 0 V to 10 V) are taken and stored. The data of $1/T^2$ vs I_C are plotted in graph display window on Page #1.
10. When the data at 10 V have been stored, click "Stop" (at the bottom-left of Page #1) to end the data collection.
11. Data display

   If the curve does not fill the graph window, click on the Scale-to-Fit tool (⬚) at the top-left corner of the graph window.
12. If you are not satisfied with the data collected in Run #N (N is probably 1), delete it by clicking "Delete Last Run" in the Control Panel, Then re-do Run #N.
13. If you are satisfied with the data collected in Run #N, click on the "Data Summary" button at the left edge of the page, right click on Run #N (N is probably 1) and rename it as "North-1".

Attention:
(1) When each of the 4 measurements is done, reset DC volt to 0 V, then click "Off".
(2) To prevent Capstone program from crash, the data fitting should be performed after all of the 4 measurements are done.

Measurement #2: Repeat Measurement #1, rename Run #N as "North-2"
Measurement #3: Repeat Measurement #1, rename Run #N as "North-3"

Measurement #4: Observation of $1/T^2$ vs I_C with mark "N" pointing south.

Repeat all steps in **Measurement #1** to observe $1/T^2$ vs I_C with mark "N" pointing south.

Attention: Observe how the bar magnet oscillates at 1 V.

Data analysis I: Summary of data from Measurements #1, #2, #3 on Graph Page #1

1. Add new columns to the Table on Page #1 to show all data from Measurements #1, #2, #3.
    (a) Click on any data row in the right-side column of the Table on Page #1 to highlight the column, then click the "Insert empty column" icon at the top-left corner of the Table to add two new columns to the right.
    (b) In each of the two new columns click "Select measurement", in the pop-up select "Period (s)".
    (c) In each of the two new columns click the "Run Name" at the top of the new column, in the pop-up select "North-2" and "North-3" respectively. The Period (T) data of "North-2" and "North-3" are displayed in the two new columns.

2. Display and fit the $1/T^2$ vs I_C curves obtained from Measurements #1, #2, #3 on Page #1
    (a) In the Tools Palette, click the Calculator icon to open the Calculator panel. Click the formula "1/T^2 =1/[Period(s), ▼]^2", then click on the small black triangle ▼ in the formula and select "All Runs".
    (b) In the graph display window on Page #1 click the Data Selection Tool (△▼) so that it is pushed in, click on ▼ and select "North-1", the $1/T^2$ vs I_C curve of "North-1" is displayed on the on Page #1.
    (c) Click on "Apply selected curve fits" in the graph Toolbar, in the pop-up select "Linear: mx + b". The fitting line along with an Info-box appear on the graph. The fitting parameters m (slope of the fitting line) and b (intercept) appear in the Info-box.
    (d) Record in Table 1 the slope and intercept values with units from the info-box.
    (e) Repeat steps (b) and (d) for "North-2" and "North-3" respectively for which step (c) is unnecessary.

3. The new Graph Page #1 should be similar to that shown in Fig. 4.

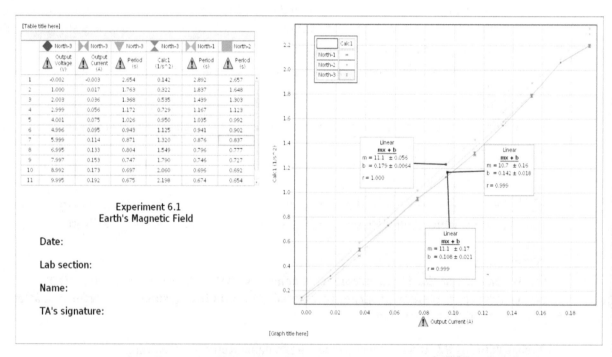

Figure 4 Summary of data from Run #1, #2, #3 on Graph Page #1

Table 1 Fitting parameters (with units) from the info-boxes in the graph display window on Page #1

Measurement	Slope	Intercept
#1 (North-1)		
#2 (North-2)		
#3 (North-3)		

Data Analysis II: Calculate constant C and B_h

1. Calculate the values of constant C in Eq. (7) using the slope values in Table 1. Record your results in Table 2.

2. Calculate the values of B_h from the values of constant C and the intercept values in Table 1. You may use $(1/T^2)_{average}$ at $I_C = 0$ in the Table on Page #1 as the intercept. Record your results in Table 2.

3. The literature value of B_h in Florida is 0.25 Gauss. Calculate the percentage error between the measured and the accepted values of B_h. Record your results in Table 2.

4. Create a Text Box on Graph Page #1 below the Table.
   (a) You may need to reduce the size of that Table on Page #1 by dragging an arrow cursor ($\Leftrightarrow$) along the diagonal from a corner of the blue frame. You can also move the Table by clicking on and dragging one of the four sides (blue frame) to where you want to place it.
   (b) Double click the Text Box icon in the Displays palette to open it on Graph Page #1.
   (c) In the Text Box type in the contents as shown in Fig. 4.
   When you put mouth cursor at any point in the Text Box a command ribbon appears at the top of Text Box. You can change the font and font size by click "A" and in the pop-up select your favor font and font size.
   (d) You may reduce the size of the Text Box by dragging an arrow cursor ($\Leftrightarrow$) along the diagonal from a corner of the blue frame. You can move the Text Box by clicking on and dragging one of the four sides (blue frame) to where you want to place it.

Table 2 Calculated values of constant C, B_h (with units) and percentage error

Measurement	C	B_h	% error
#1			
#2			
#3			

Work to be done:

1. Print Graph Page #1 (similar to Fig. 4) as a part of your lab report on Experiment 6.1.
   Click on "File" at the top-left" corner, in the pop-up select "Print Page Setup" and "Landscape", then print it out.
2. Let your TA check your Page #1. If it is OK, your TA will sign it.
   Caution: Make sure that your printed Page #1 is OK before you close the Capstone. All the data collected in Experiment 6.1 will be lost when the Capstone is closed!
3. Close PASCO Capstone by clicking the red-cross at the upper corner, then select "discard".
4. Turn off PASCO 850 interface, leave the computer on.
5. Clean up your bench and setup apparatus of Experiment 6.2.

Lab report on Experiments 6.1

1. Your lab report should be in the required format described in the "Introduction" of the lab manual.
2. Graph Page #1 (similar to Fig. 4) should be included in your lab report.
3. Tables 1 and 2 should be included in your lab report. It is required that the answers or solutions to the questions (see below) should be included in your lab report.
4. You can tear those pages out of the lab manual as a part of your lab report, which contain measured (raw) data and analyzed data, answers to questions. The data sheets must be checked and signed by your lab TA.

Questions

1. Magnetic field ($\vec{B}$) exerts a torque ($\vec{\tau}$) on a magnetic moment ($\vec{m}$) (such as a magnet bar) to align the magnetic moment along the $\vec{B}$ – field direction.

    In general, the vector torque is defined as: $\vec{\tau} = \vec{m} \times \vec{B}$, with a magnitude of $\tau = |\vec{\tau}| = m B \sin \theta$, where θ is the angle between $\vec{m}$ and $\vec{B}$ as shown in Fig. 1.

    However, in Eq. (1) the magnitude of $\vec{\tau}$ is defined as $\tau = -m B \sin \theta$. What does the minus sign mean?

2. What are the SI units of the fitting slope and intercept in Fig. 4 and Table 1? (Show your work).

3. What is the SI units of the constant C in Table 2? (Show your work).

4. Why does the bar magnetic rotate rather than oscillate at DC Volt = 1 V in Measurement #4?

Experiment 6.2
Magnetic Field of a Solenoid

Purpose
Measure the magnetic field inside a solenoid.

Equipment
A solenoid (2920 turns, 9.8 cm long, 3 cm in diameter), 2-axis magnetic field sensor (PS-2162) with a home-made guide tube and a home-made base, optical bench 60 cm, PASCO 850 interface, computer.

Theory
The magnetic field inside a very long solenoid is given by:
$$B = \mu_0 n I,$$
(1)

where $\mu_0 = 4\pi \times 10^{-7}$ $Tm/A = 4\pi \times 10^{-3}$ $Gauss\ m/A$, I is the current (A) through the solenoid, n is the number of turns of wire per unit length (# /m) of the solenoid.

Apparatus setup
Caution: Turn off the power of PASCO 850 interface before you make connections. Don't let the two output terminals from the output ports on 850 interface touch each other anytime.

1. 2-axis magnetic field sensor (PS-2162) (Fig. 1).
   The sensor measures the magnetic field strength in the **Axial** and **Perpendicular (radial)** directions simultaneously. Two white dots on the probe mark the positions of the sensing elements.

2. Insert the probe of the magnetic field sensor into the solenoid (Fig. 2) with a home-made guide tube and a home-made base. Position the sensing elements (marked by white dots) at the center of the solenoid and orient the sensor with the field lines.

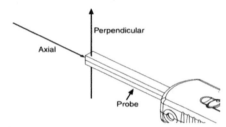

Figure 1   2-axis magnetic field sensor

3. Plug the cable of the Magnetic field sensor into PASPORT inputs on 850 Interface (Fig. 2).

4. Use one red and one black banana leads to connect the solenoid and the Output Ports (banana jacks) on PASCO 850 Interface which provides DC power for the solenoid. (Fig. 2).

5. A built-in current sensor of the 850 Interface is used to measure the current flowing through the solenoid.

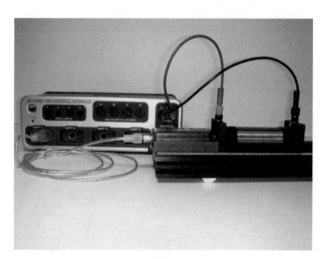

Figure 2 Apparatus setup

Software setup

1. Turn on PASCO 850 Interface.
2. Start PASCO Capstone by double clicking the symbol   on computer. The Workbook Page appears.
3. Hardware Setup on the Workbook Page
   (a) In the Tools Palette, click on the "Hardware Setup" icon, the Hardware Setup panel appears with a picture of the PASCO 850 interface along with a Magnetic Field Sensor icon below the PASPORT inputs.
   (b) Click the Output Ports on the upper-right corner of the picture of 850 Interface, in the pop-up select the "Output Voltage-Current Sensor", the sensor icon [⚡] appears on the top.
   (c) Click on the "Hardware Setup" icon to close the Hardware Setup panel.
4. Setup Table & Graph Page #1 on the Workbook page:
   (a) Click on the Table & Graph icon on the Workbook page, a Table and a Graph display window appear on Page #1.
   (b) Setup the Table on Page #1
      • In the 1st column of the Table, click "Select measurement", in the pop-up select "Output Voltage (V)".
      • In the 2nd column of the Table, click "Select measurement", in the pop-up select "Output Current (A)".
      • Click the "Insert empty column" icon at the top-left corner of the Table to add a 3rd column, and click "Select measurement", in the pop-up select "Magnetic Field Strength (parallel) (T)", then rename it as "B-axial (Gauss).
      • Click the "Insert empty column" icon at the top-left corner of the Table to add a 4th column, and click "Select measurement", in the pop-up select "Magnetic Field Strength (perpendicular) (T)", then rename it as "B-radial (Gauss).
   (c) Set up the Graph display window on Page #1
      • Click the x-axis label "Select measurement", in the pop-up select "Output Current (A)".
      • Click the y-axis label "Select measurement", in the pop-up select "B-axial (gauss)".

Data collection

1. In the Controls Palette (at the bottom of the Workbook Page) select "Keep Mode".
2. In the Controls Palette set sampling rate for Output Voltage-Current Sensor to 200 Hz.
3. Setup Signal Generator on the Workbook Page
   (a) In the Tools Palette, click "Signal Generator" button to open the Signal Generator panel.
   (b) Click on "850 Output 1" and make the following selections:
      • Wave form: DC
      • DC Volt: 0 V
      • Auto
4. The left side of Page #1 is blocked by the "Signal generator panel". Click the red pin ⚲ at the upper-right corner of the Tools panel, then the full Page #1 is displayed on the right side of the "Signal generator panel".
5. Press the "Tare" button on the magnetic field sensor to zero it before data collection.
6. Insert the probe of the magnetic field sensor into the center of the solenoid.
7. Click "Preview" (at the bottom-left of Page #1), the data of voltage, current, B-axial and B-radial appear in the Table on Page #1.
8. When data are stable, click "Keep Sample" (at the bottom-left of Page #1) to store the data.
9. Increase the DC voltage 1 V each time and repeat step 8. The final output voltage from the DC power supply is 15 V, so total 16 data points (from 0 V to 15 V) are taken and stored. The data of B-axial vs current I are plotted in graph display window on Page #1.

10. When the data at 15 V have been stored, click "Stop" (at the bottom-left of Page #1) to end the data collection.

11. Data display: if the curve does not fill the graph window, click on the Scale-to-Fit tool ( ) at the top-left corner of the graph window.

12. If you are not satisfied with the data collected in Run #1, delete it by clicking "Delete Last Run" in the Control Panel, Then re-do Run #1.

13. On Signal Generator panel, reset DC volt to 0 V and click "Off".

Data analysis

1. Calculate B-theory
   (a) Highlight column 2 (Output Current) to make it active, then create a new calculation column next to it and rename the new calculation as B-theory.
   (b) The complete the expression for the theoretical calculation using Eq. (1) should be:

   "B-theory"=[Output Current(A), Run #1]* μ_0 *n with units gauss.

   (c) In the graph display window click "Add new y-axis", a new y-axis appears on the right side of the graph.
   (d) Click the right-side y-axis label, in the pup-up select "B-theory (Gauss)". The B-theory data are plotted along with the plotted data of B-axial (Gauss).

2. Display B-radial
   (a) In the graph display window click "Add new y-axis", a new y-axis appears on the left side of the graph.
   (b) Click on the new y-axis label, in the pop-up select "B-radial (gauss)".
   (c) If all the B-radial data are negative, click on the y-axis label "B-radial", in the pop-up select "Quick Calc" and "$-B_\perp$".
   (d) Same y-scale must be used for "B-axial", "B-theory" and "B-radial" in order to compare them.

3. Linear fitting of B-axial and B-theory curves
   (a) Click the B-axial curve to make it active, then click on "Apply selected curve fits" in the graph Toolbar, in the pop-up select "Linear: mx + b". The fitting line along with an Info-box appear on the graph. The fitting parameters m (slope of the fitting line) and b (intercept) appear in the Info-box. The slope can be used to calculate the total number of turns of the solenoid.
   (b) Repeat step (a) for the linear fitting of B-theory curve.

4. Setup the Text Box as shown in Fig. 3 using similar steps described in the Data Analysis II of Experiment 6.1
   Now the Page #1 should be similar to Fig. 3.

Work to be done:

1. Print Graph Pages #1 (similar to Fig. 3) as a part of your lab report on Experiment 6.2.
   Click on "File" at the top-left" corner, in the pop-up select "Print Page Setup" and "Landscape", then print it out.

2. Let your TA check your Graph Pages #1. If it is OK, your TA will sign it.
   Caution: Make sure that your printed Graph Pages #1 is OK before you close the Capstone. All the data collected in Experiment 6.2 will be lost when the Capstone is closed!

3. Close PASCO Capstone by clicking the red-cross at the upper corner, then select "discard".

4. Turn off PASCO 850 interface and shut down the computer.

5. Clean up your bench.

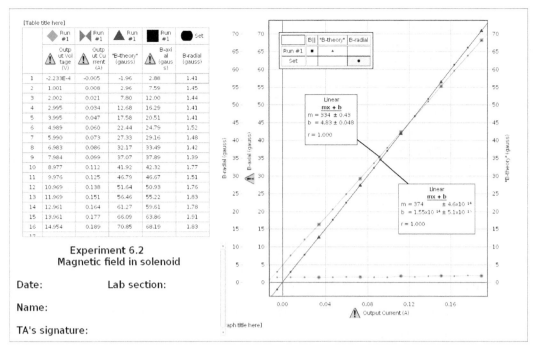

[Table title here]

	Output Voltage (V)	Output Current (A)	"B-theory" (gauss)	B-axial (gauss)	B-radial (gauss)
1	-2.233E-4	-0.005	-1.96	2.88	1.41
2	1.001	0.008	2.96	7.59	1.45
3	2.002	0.021	7.80	12.00	1.44
4	2.995	0.034	12.68	16.29	1.41
5	3.995	0.047	17.58	20.51	1.41
6	4.989	0.060	22.44	24.79	1.52
7	5.990	0.073	27.33	29.16	1.48
8	6.983	0.086	32.17	33.49	1.42
9	7.984	0.099	37.07	37.89	1.39
10	8.977	0.112	41.92	42.32	1.77
11	9.976	0.125	46.79	46.67	1.51
12	10.969	0.138	51.64	50.93	1.76
13	11.969	0.151	56.46	55.22	1.83
14	12.961	0.164	61.27	59.61	1.78
15	13.961	0.177	66.09	63.86	1.91
16	14.954	0.189	70.85	68.19	1.83

Experiment 6.2
Magnetic field in solenoid

Date: Lab section:

Name:

TA's signature:

Linear
mx + b
m = 334 ± 0.43
b = 4.83 ± 0.048
r = 1.000

Linear
mx + b
m = 374   ± 4.6x10⁻¹⁴
b = 1.55x10⁻¹⁴ ± 5.1x10⁻¹⁵
r = 1.000

Output Current (A)

Figure 3   Magnetic field in solenoid ---- comparison of B-axial with B-theory

Lab report on Experiments 6.2

1. Your lab report should be in the required format described in the "Introduction" of the lab manual.
2. Graph Page #1 (similar to Fig. 3) should be included in your lab report.
3. It is required that the answers or solutions to the questions (see below) should be included in your lab report.
4. You can tear those pages out of the lab manual as a part of your lab report, which contain measured (raw) data and analyzed data, answers to questions. The data sheets must be checked and signed by your lab TA.

Questions

1. Is the magnet field B-axial positive or negative? Why? If it is negative, how to make it positive?

2. What are the units of the fitting slope and intercept in Fig. 3?

3. The total number of turns of the solenoid (N_{exp}) can be found from the fitting slope of the B-axial curve shown in the graph window of Fig. 3. Compare N_{exp} with the manufacturer value N which can be found in this manual.

Experiment 7.1
Induced emf: magnet passes through a solenoid

Purpose

Measure the electromotive force (emf) induced by dropping a magnet through the center of a solenoid.

Equipment

A solenoid, a pair of magnets (alnico), PASCO 850 interface, voltage sensor (PASCO UI-5100), computer, a short cylindrical bar magnet and a copper pipe, a set of "jumping ring".

Theory

When a magnet is passing through a coil there is a changing in magnetic flux through the coil, which induces an electromotive force (emf) in the coil. According to Faraday's law of induction:

$$\varepsilon = -N \frac{\Delta \Phi}{\Delta t} \qquad (1)$$

Where ε is the induced emf, N is the number of turns of wire in the coil, and $\frac{\Delta \Phi}{\Delta t}$ is the rate of change of the magnetic flux through each turn of the coil.

The negative sign in Eq. (1) is used to determine the polarity of the induced emf or the direction of the induced current. An alternative method for determining the direction of the induced current is the **Lenz's law**: "The direction of any magnetic induction effect is such as to oppose the cause of the effect".

Incoming flux, incoming peak, outgoing flux and outgoing peak
When the magnet is entering the coil the magnetic flux through the coil increases with time. This **increasing flux** is called incoming flux and the emf induced by the incoming flux versus time is displayed as a peak called the **incoming peak**.

When the magnet is leaving the coil the magnetic flux through the coil decreases with time. This decreasing flux is called **outgoing flux** and the emf induced by the outgoing flux versus time is displayed as a peak called the **outgoing peak**.

The incoming and outgoing peaks can be displayed as either positive or negative peaks depending on the connections of the voltage sensor to the coil. After the connections of the voltage sensor to the coil are made the polarities (positive or negative) of the incoming and outgoing peaks are determined by the orientations of the magnet.

In this experiment, the voltage sensor measures the voltage (emf) induced by dropping a magnet through the center of a coil. The program records and displays the induced emf (voltage) versus time, and calculates the areas under the incoming and outgoing peaks which represent the change in total flux through the coil since $\varepsilon \ \Delta t \propto \Delta \Phi$.

Experiment apparatus

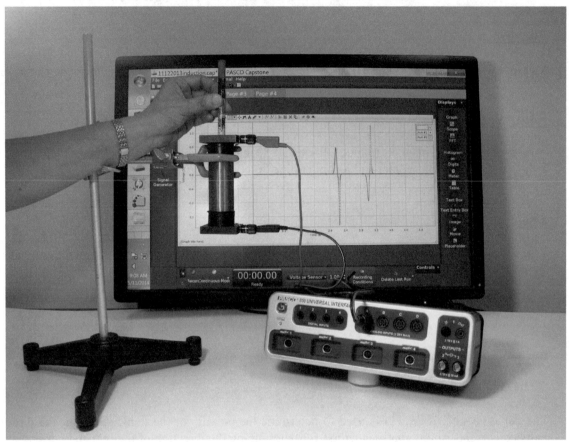

Figure 1   Experiment Apparatus

Measurements to be done

Measurement #1: emf (V) versus time (t). The emf is induced by dropping a magnet
(with the **N**-pole downward) into the center of the solenoid.

Measurement #2:   Same as **Data Set 1** but with the **S**-pole of the magnet downward.

Measurement #3:   emf (V) versus time (t). The emf is induced by dropping **two magnets**
(in series as shown, and with the **N**-pole downward) into the center of the solenoid.

N	S	N	S

Measurement #4:   Same as **Data Set 3** but with the **S**-pole of the magnets downward.

Attention: the initial position of the bottom of the magnets should be the same for all four measurements:
0.5 mm above the top center of the solenoid.

Preparation

The sketch of PASCO 850 interface is given in the "Introduction" of this lab manual.

Caution: **Turn off the power of PASCO 850 interface before you make connections. Don't let the two output terminals from the Output Ports on 850 interface touch each other anytime**.

1. Use voltage sensor to connect the solenoid and Port A of the Analog Inputs on 850 Interface (Fig. 1).
2. Turn on PASCO 850 Interface and computer. Log in computer using Password: phy2049.
3. Start PASCO Capstone by double clicking the symbol on computer.
   The Workbook Page appears.
4. Hardware Setup on the Workbook Page
   (a) In the Tools Palette, click on the "Hardware Setup" icon, the Hardware Setup panel appears with a picture of the PASCO 850 Interface along with a 'voltage sensor' icon on the top of ports A on the Analog Inputs.
   Attention: If the 'voltage sensor' icon does not show up automatically, click Port A on the Analog Inputs (in the PASCO 850 Interface picture), a drop down menu of sensors appears. Type in "v", in the pop-up select "voltage sensor", then a 'voltage sensor' icon appears.
   (b) In the Controls Palette (at the bottom of the Workbook Page) select "Continuous Mode" and set sampling rate to 1 kHz.
   (c) Click on "Hardware Setup" icon to close the Hardware Setup panel.

Measurement #1 One magnet with N-pole downward (denoted **1mNd**)

1. Setup Graph Display Page #1
   (a) Double click on the graph icon in the Displays palette, a graph display appears.
   (b) Click on the y-axis label and in the pop-up select Ch A (V).
   (c) The x-axis label should be time (s).
2. Hold a single magnet vertically with N-pole downward and keep its bottom 0.5 mm above the top center of the solenoid.
3. Click "Record" button in the Controls palette on the lower-left corner of the Workbook Page to start data collection.
4. About 2 seconds after the "Record" button is clicked drop the magnet into the center of the solenoid vertically. A V(t) curve (with two peaks) appears on the graph display.
5. Click Stop button to end data collection.
6. Examine the V(t) curve:

   (a) If the V(t) curve does not fill the graph, click on the Scale-to-Fit tool (⊡) at the upper left of the Graph Toolbar.
   (b) You can enlarge or reduce the curve size vertically by changing the y-axis scale: move the hand cursor above a number on the y-axis scale and when the hand cursor changes to a vertical ⇕ cursor, click and drag the ⇕ cursor upward (for enlarging) or downward (for reducing) along the y-axis.
   (c) Similarly, you can change the x-axis scale by dragging the horizontal cursor ⇔ to right (for enlarging) or to left (for reducing) along the x-axis.
   (d) Label coordinates for each of the two peaks

   Click on the Coordinates Tool (⊹) from the graph Toolbar, a square box appears, click on the box and drag it until it is directly above the peak to be labeled (above it, not on the line – you should see an arrow pointing down to the peak). Release the cursor and it will snap to the peak. The two numbers in the info box are [time, voltage]. Repeat this step for the other peak.

7. If you are not satisfied with the V(t) curve, click on the "Delete Last Run" button in the Controls Palette, and repeat measurement #1.

8. If you are satisfied with the V(t) curve, click on the "Data Summary" button at the left edge of the page, double click on Run #1 and re-label it "**1mNd**". Click on the Data Summary button to close the Data Summary panel.

Measurement #2:   One magnet with S pole downward (denoted **1mSd**)

Repeat steps 2 to 8 in "measurement #1" to perform Measurement #2 with the following changes:
1. In step 2 change N-pole to S-pole.
2. In step 4 change "About 2 seconds" to "About 3 seconds".
   Note: In order to display all four V(t) curves from 4 measurements on the same graph page without overlapping, a time delay in step 4 is needed.
3. In step 8 change "click on Run #1 and re-label it '**1mNd**'" to "click on Run #2 and re-label it '**1mSd**'".

Simultaneous viewing of multiple data sets

Now you have two V(t) curves: **1mNd** and **1mSd**.

1. If you want to display one curve on the graph, click on the Data Selection Tool (▲▼) so that it is pushed in and then click on the small black triangle and select one of the two curves that you want to display.
2. You can display the two V(t) curves simultaneously on the same graph page by selecting both of them using the Selection Tool (▲▼).
3. There is a legend (an information table) on the upper right corner of the graph page which lists the all the selected data sets displayed on the same graph. If you want to do something (such as peak labeling or color change) on one of the curves, click on the name of the curve in the legend, the curve will be highlighted (i.e., becomes active). The legend can be moved around by clicking on and dragging the top empty box or one of the four sides (frame) of the legend to where you want to place it.

Measurement #3:   Two magnets in series with N pole downward (denoted **2mNd**)

Repeat steps 2 to 8 in "measurement #1" to perform Measurement #3 with the following changes:
1. In step 2 change "single magnet" to "two magnets in series".
2. In step 4 change "About 2 seconds" to "About 6 seconds".
3. In step 8 change "click on Run #1 and re-label it '**1mNd**'" to "click on Run #3 and re-label it '**2mNd**'".

Measurement #4:   Two magnets in series with S pole downward (denoted **2mSd**)

Repeat steps 2 to 8 in "measurement #1" to perform Measurement #4 with the following changes:
1. In step 2 change "single magnet" to "two magnets in series".
2. In step 4 change "About 2 seconds" to "About 7 seconds".
3. In step 8 change "click on Run #1 and re-label it '**1mNd**'" to "click on Run #4 and re-label it '**2mSd**'".

Data Analysis

1. Use the display method described in "**Simultaneous viewing of multiple data sets**" to display all four V(t) curves on the same graph page for Data Analysis.
2. Record the incoming and outgoing voltages in Table 1.
3. Create a Text Box on the Graph Page #1.
   (a) Double click the Text Box icon in the Displays palette to open it on Page #1.

(b) You may reduce the size of the Text Box by dragging an arrow cursor ($\Leftrightarrow$) along the diagonal from a corner of the blue frame.

(c) You can move the Text Box around by clicking on and dragging one of the four sides (blue frame) to where you want to place it.

(d) Setup the Text Box as shown in Fig. 2 and type in content similar to that in Fig. 2.

When you put mouth cursor at any point in the Text Box a command ribbon appears at the top of Text Box. You can change the font and font size by click "A" and in the pop-up select your favor font and font size.

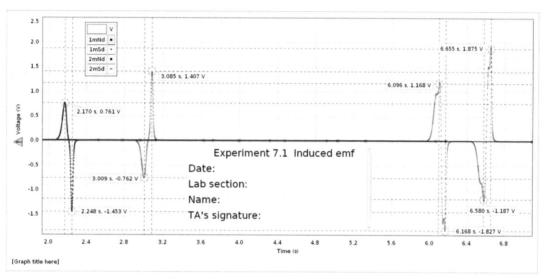

Figure 2 Analyzed data of Experiment 7.1

Table 1 Incoming and outgoing voltages from the 4 measurements shown in Fig. 2.

Measurement	Incoming voltage (V)	Outgoing voltage (V)
#1: 1mNd		
#2: 1mSd		
#3: 2mNd		
#4: 2mSd		

Work to be done:

1. Print Graph Page #1 (similar to Fig. 2) as a part of your lab report on Experiment 7.1.

   Click on "File" at the top-left" corner, in the pop-up select "Print Page Setup" and "Landscape", then print it out.

2. Let your TA check your Page #1, if it is OK your TA will sign it.

   Caution: Make sure that your printed Page #1 is OK before you close the Capstone. All the data collected in Experiment 7.1 will be lost when the Capstone is closed!

3. Close PASCO Capstone by clicking the red-cross at the upper corner, then select "discard".

4. Turn off PASCO 850 interface, leave the computer on.

5. Clean up your bench and setup apparatus of Experiment 7.2.

Lab report on Experiment 7.1

1. Your lab report should be in the required format described in the "Introduction" of the lab manual.
2. Graph Page #1 (similar to Fig. 2) should be included in your lab report.
3. Table 1 should be included in your lab report.
4. It is required that the answers or solutions to the questions (see below) should be included in your lab report.
5. You can tear those pages out of the lab manual as a part of your lab report, which contain measured (raw) data and analyzed data, answers to questions. The data sheets must be checked and signed by your lab TA.

Questions

1. The four V(t) curves from four measurements show that the incoming and outgoing peaks are opposite in direction. Why?

2. Explain why the magnitude of the incoming peak is less than the outgoing peak as shown in the four V(t) curves and in the Table on the graph page.

3. Explain why the magnitudes of both the incoming and outgoing peaks in **2mNd** curve are greater than that in **1mNd** curve (see the values in the Table).

4. Use Lenz's law to determine whether the induced current in the resistor R is from A to B or from B to A in the 4 configurations in Fig. 3. You should label the direction of $\vec{B}_{source}$, the sign of $\Delta\Phi_B / \Delta t$, the direction of $\vec{B}_{induced}$ and the direction of $I_{induced}$ in the coil and R for each of the 4 cases.

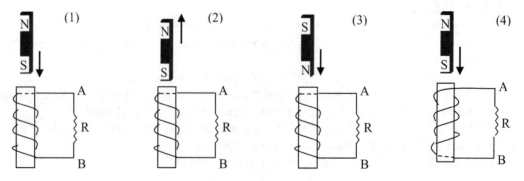

Figure 3 Four configurations

5. Drop a short cylindrical bar magnet down a vertical copper pipe (Fig. 4).

    A short cylindrical bar magnet is dropped down a vertical copper pipe of slightly large diameter, about 1.0 m long. It takes several seconds to merge at the bottom, whereas it takes a fraction of a second to make the trip in the air. Why?

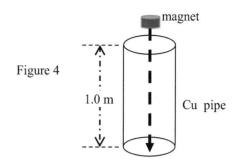

Figure 4

6. The "jumping ring" (Fig. 5).

    If you wind a solenoid coil around an iron core (the iron is there to beef up the magnetic field). Place an aluminum ring on top, and press the switch button, the ring will jump several feet in the air. Why?

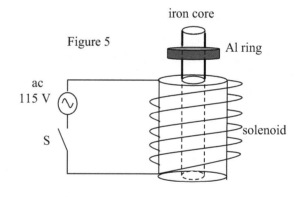

Figure 5

Experiment 7.2
Transformer

Purpose
Learn how a transformer converts ac voltage.

Equipment
Two solenoids (with total turns of 2920 and 235 respectively), iron bar (SE-8653), Cu, Al and plastic bar, PASCO 850 interface, two voltage sensors (PASCO UI-5100), computer, one digital multimeter (DMM).

Theory
A transformer is an electrical device that transfers energy between two circuits through electromagnetic induction. A transformer may be used as a safe and efficient voltage converter to change the ac voltage at its input to a higher or lower voltage at its output. Other uses include current conversion, isolation with or without changing voltage and impedance conversion.

The key components of the transformer are two coils, electrically insulated from each other but wound on the same core as shown in Fig. 1. The core is made of a material with large permeability to keep the magnetic field lines due to a current in one coil almost completely within the core. Hence almost all of these field lines pass through the other coil, maximizing the mutual inductance of the two coils. The coil to which power is supplied is called the **primary coil** with N_p turns; the coil from which power is delivered is called the **secondary coil** with N_s turns.

The ac source causes an alternating current in the primary, which sets up an alternating flux in the core; this induces an emf in each coil, in accordance with Faraday's law. The induced emf in the secondary gives rise to an alternating current in the secondary, and delivers energy to the device to which the secondary is connected. Current and emf in the secondary coil have the same frequency as the ac source.

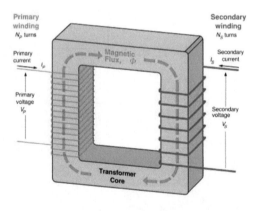

Figure 1
Schematic diagram of
an idealized transformer

According to Faraday's law of induction, the induced emf (voltage) is proportional to the rate of change in magnetic flux through the coil ($d\phi/dt$) and the number of turns (N) in the coil:

$$\varepsilon = -N \frac{\Delta \Phi}{\Delta t} \tag{1}$$

Since the rate of change in flux through both coils is the same, the ratio of the emfs (voltages) in the coils is equal to the ratio of the numbers of turns (N) in the coils:

$$\because \quad \varepsilon_s = -N_s \frac{\Delta \Phi}{\Delta t} \qquad\qquad \varepsilon_p = -N_p \frac{\Delta \Phi}{\Delta t}$$

For an idealized transformer, $\varepsilon_p \simeq V_p$, $\varepsilon_s \simeq V_s$, $\qquad \therefore \quad \dfrac{\varepsilon_s}{\varepsilon_p} = \dfrac{N_s}{N_p} \simeq \dfrac{V_s}{V_p}$ (2)

A transformer can be used to increase or decrease ac voltages.

If $N_s > N_p$, then $V_s > V_p$, --- step-up transformer;

If $N_s < N_p$, then $V_s < V_p$, --- step-down transformer.

Experiment apparatus

Figure 2 shows the experiment apparatus.

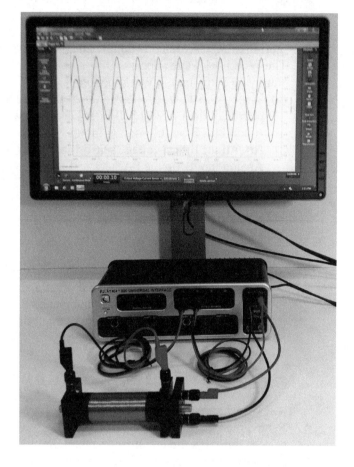

Figure 2   Experiment Apparatus

Measurements and observations to be made

Measurement #1: Step-up transformer with an iron bar inside the inner solenoid. Measure the peak value V_s and frequency f_s and observe the phase shift of the secondary signal.

Measurement #2: Step-down transformer with an iron bar inside the inner solenoid. Measure the peak value V_s and frequency f_s and observe the phase shift of the secondary signal.

Observation #1: Same as Measurement #1, but take out the iron bar from the inner solenoid to observe the changes in the secondary signal including its peak value, frequency and phase shift.

Observation #2: Same as Measurement #1, but put a Cu bar, or an Al bar, or a plastic bar inside the inner solenoid respectively to observe the changes in the secondary signal including its peak value, frequency and phase shift.

Measurement #1: Step-up transformer with an iron bar inside the inner solenoid. Measure the peak value V_s and frequency f_s and observe the phase shift of the secondary signal.

1. Make the following connections referring to Fig. 2:

    Caution: Turn off the power of PASCO 850 interface before you make connections. Don't let the two output terminals from the output ports on 850 interface touch each other anytime.

    (a) Use one red and one black banana leads to connect the inner solenoid (primary coil) and the Output Ports (banana jacks) for signal generator on 850 interface.

    (b) Use a voltage sensor to connect the outer solenoid (secondary coil) and Port A of the Analog Inputs on 850 Interface.

    (c) Use a voltage sensor to connect Port B of the Analog Inputs and the Output Ports (banana jacks) for signal generator on 850 interface.

    (d) Turn on 850 interface.

2. Start PASCO Capstone by double clicking the symbol on computer.
    The Workbook Page appears.

3. Hardware Setup on the Workbook Page

    (a) In the Tools Palette, click on the "Hardware Setup" icon, the Hardware Setup panel appears with a picture of the PASCO 850 Interface along with twp 'voltage sensor' icons ⊡ on the top of Port A and Port B on the Analog Inputs.

    Attention: If the 'voltage sensor' icon does not show up automatically, click Port A (or B) on the Analog Inputs (in the PASCO 850 Interface picture), a drop down menu of sensors appears. Type in "v", in the pop-up select "voltage sensor", then a 'voltage sensor' icon appears.

    (b) In the Controls Palette (at the bottom of the Workbook Page)
    - select "Continuous Mode",
    - set sampling rate to 200 kHz,
    - click "Recording conditions", select "Stop condition"/"Time base" and set to 0.1 s.

    (c) Click on "Hardware Setup" icon to close the Hardware Setup panel.

4. Setup Signal Generator on the Workbook Page

    (a) In the Tools Palette, click "Signal Generator" button to open the Signal Generator panel.

    (b) Click on "850 Output 1" and make the following selections:
    - Wave form: Sine
    - Sweep: Off
    - Frequency: 100 Hz
    - Amplitude: 0.2 V
    - Auto

5. Setup graph Page #1 on the Workbook page:

    (a) Double click the graph icon in the Displays palette, to open graph Page #1 on Workbook page.
    Note: If the left side of the graph Page #1 is blocked by any Tools panel such as "Hardware Setup panel" or Signal generator panel", click on the red pin ⚲ at the upper-right corner of the Tools panel, then the full graph Page #1 is displayed on the right side of the Tools panel.

    (b) Click on the y-axis label and in the pop-up select "voltage Ch B (V)" --- the voltage $v_p(t)$ across the inner solenoid (primary coil).

    (c) Click "Add new y-axis to active plot area" on the top of the graph Tool bar, the new y-axis is displayed on the right side of graph Page #1. Set the new y-axis to "voltage Ch A (V)" --- the voltage $v_s(t)$ across the outer solenoid (secondary coil).

6. Data collection
   (a) Click the Record button at the left end of the Controls palette to start data collection which will be automatically stopped at 0.1 s. Ten cycles of $\upsilon_p(t)$ and $\upsilon_s(t)$ curves are displayed.

   - If the $\upsilon_p(t)$ and $\upsilon_s(t)$ curves do not fill the graph, click on the Scale-to-Fit tool ( ) at the upper left of the Graph Toolbar.
   - **Attention:** If the $\upsilon_p(t)$ and $\upsilon_s(t)$ curves are out of phase (with 180^0 phase difference), exchange the red and black banana leads on the secondary coil to make the two curves in phase (with 0^0 or 360^0 phase difference), then re-take the data.
   - There is a legend (an information table) on the upper right corner of the graph Page #1 which indicates that both V_0 (i.e., V_p) and V (i.e., V_s) curves are displayed on the same graph. If you want to do something (such as peak labeling or color change) on one of the curves, click on the name of the curve in the legend, the curve will be highlighted (i.e., becomes active).

   (b) The amplitude of $\upsilon_p(t)$ is much less than that of $\upsilon_s(t)$ because of the step-up transformer. For clarity, enlarge the $\upsilon_p(t)$ curve size vertically by changing the left side y-axis scale: click on V_0 in the legend to highlight the $\upsilon_p(t)$ curve, then move the hand cursor above a number on the right side y-axis scale, when the hand cursor changes to a vertical $\Updownarrow$ cursor, click and drag the $\Updownarrow$ cursor upward to enlarge the vertical size of the $\upsilon_p(t)$ curve. This action won't change the peak value.

   (c) If you are not satisfied with the $\upsilon_p(t)$ and $\upsilon_s(t)$ curves, click on the "Delete Last Run" button in the Controls Palette, and repeat measurement.

   (d) If you are satisfied with the $\upsilon_p(t)$ and $\upsilon_s(t)$ curves, click on the "Data Summary" button at the left edge of the page, right click on Run # 1 and rename it "Step-up with Fe bar". Click on the Data Summary button to close the Data Summary panel.

   (e) Reset the amplitude to 0 V on the Signal Generator panel and click "off".

7. Peak labeling (shown in Fig. 3)
   (a) Use the Coordinates Tool (called Smart Cursor) ( ) to label the 5th and 9th (denoted peak 2) peaks of the $\upsilon_s(t)$ curve, because the amplitudes of the left 4 peaks are not constant.

   (b) The two numbers in the info box are [time (t), voltage (V)]. The info box at the 5th peak (denoted peak 1) gives $[[t_1, V_{s1}]$ and the info box at the 9th peak (denoted peak 2) gives $[t_2, V_{s2}]$. Record these data in Table 1.

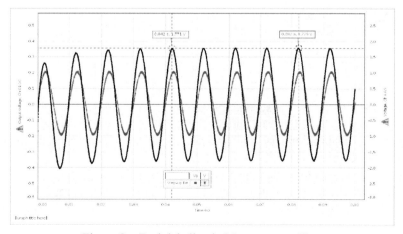

Figure 3   Peak labeling in Measurement #1

69

Table 1 Measured values (with units) of the secondary signal
Note: the outer solenoid has total 2920 turns and the inner solenoid has total 235 turns.

Transformer	Primary coil		Secondary coil								V_s / V_p
	V_p	f_p	t_1	V_{s1}	t_2	V_{s2}	# of cycles	f_s	V_s (ave)		
Step-up	0.2	100									
Step-down	5.0	100									

Measurement #2: Step-down transformer with an iron bar inside the inner solenoid. Measure the peak value V_s and frequency f_s and observe the phase shift of the secondary signal.

1. Turn off 850 interface.
2. In this measurement, the outer solenoid is the primary coil and the inner solenoid is the secondary coil.
   (a) Use one red and one black banana leads to connect the outer solenoid (primary coil) and the Output Ports (banana jacks) for signal generator on 850 interface.
   (b) Use a voltage sensor to connect the inner solenoid (secondary coil) and Port A of the Analog Inputs on 850 Interface.
   (c) Use a voltage sensor to connect Port B of the Analog Inputs and the Output Ports (banana jacks) for signal generator on 850 interface.
3. Turn on 850 interface.
4. Open the Signal Generator panel, select 850 Output 1, set Amplitude: 5.0 V, keep all other settings unchanged.
5. Perform Measurement #2 using similar steps as used in Measurement #1.
   Note: In Measurement #2 the amplitude of $\upsilon_s(t)$ is much less than that of $\upsilon_p(t)$ because of the step-down transformer. For clarity, you need to enlarge the size of the $\upsilon_s(t)$ curve vertically by changing the right side y-axis scale.
6. If you are satisfied with the $\upsilon_p(t)$ and $\upsilon_s(t)$ curves, click on the "Data Summary" button and rename run #2 as "Step-down with Fe bar", record the measured values and the observed result in Table 1.
   Caution: when you use "Delete Last Run" function to delete the unsatisfied data in Measurement #2 make sure that you do not delete the data from Measurement #1.

Observations: Same as Measurement #1, but take out the iron bar from the inner solenoid to observe the changes in the secondary signal including its peak value, frequency and phase shift.

1. All the connections and parameter settings are the same as in Measurement #1.
2. Take out the iron bar from the inner solenoid.
3. Click the Record button the $\upsilon_p(t)$ and $\upsilon_s(t)$ curves are displayed.
4. Observe the changes in the secondary signal including its peak value, frequency and phase shift.
   Note: Do not exchange the red and black banana leads on the secondary coil in all the observations.
5. Click on the "Data Summary" button and rename run #3 as "Step-up No any bar".
6. Record the observation results in Table 2.
7. Put an Al bar inside the inner solenoid, repeat steps 4 to 6.
8. Take out the Al bar and put a Cu bar inside the inner solenoid, repeat steps 4 to 6.
9. **Reset the amplitude to 0 V on the Signal Generator panel and click "off".**

Table 2 Observation results

	Estimate $V_s/V_p = ?$	Is $f_s = f_p$?	Is there a phase shift?
With no any bar			
With Al bar			
With Cu bar			

Data analysis:

1. Display on Page #1 the two sets of data from Measurements #1 and #2
   (a) Add a graph on Page #1 with the same x- and y-axis labels to display the data of both "Step-up with Fe bar" and "Step-down with Fe bar".
   (b) On each graph click the Data Selection Tool (△▼) so that it is pushed out, click on the small black triangle then select "Step-up with Fe bar" on graph 1 and "Step-down with Fe bar" on graph 2.
   (c) Repeat step 5 in "Measurement #1" to label peaks displayed on graph 2 of Page #1, and record the data from the info-boxes in Table 1.
   (c) Setup the Text Box as shown in Fig. 4.
2. Display on Page #2 the three sets data from "Observations".
   (a) Open Page #2 by clicking "Add Page" on the left of "Page #1". Setup Page #2 in the same way as for Page #1 in Measurement #1.
   (b) Setup three graphs on Page #2 with the same x- and y-axis labels as used in the graphs on Page #1.
   (c) On each graph click the Data Selection Tool (△▼) so that it is pushed out, click on the small black triangle then select one of the three data sets ("Step-up no any bar", "Step-up with Cu bar" and "Step-up with Al bar") on each of the three graphs respectively.
   (d) Setup the Text Box as shown in Fig. 5.

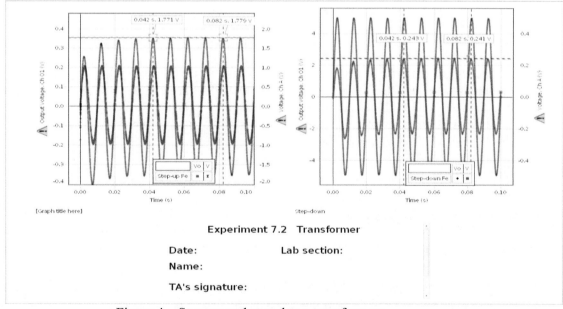

Figure 4   Step-up and step-down transformers

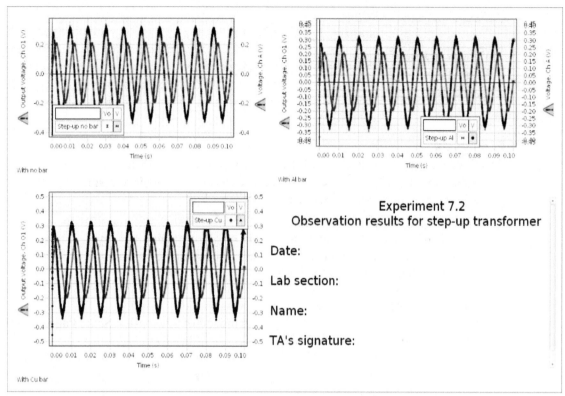

Figure 5   Observation results for step-up transformer

Work to be done:

1. Print Graph Page #1 (similar to Fig. 4) and Graph Page #3 (similar to Fig. 5) as a part of your lab report on Experiment 7.2.

   Click on "File" at the top-left" corner, in the pop-up select "Print Page Setup" and "Landscape", then print it out.

2. Let your TA check your Graph Pages #1 and #3, if they are OK your TA will sign them.

   Caution: Make sure that your printed Pages #1 and #3 are OK before you close the Capstone. All the data collected in Experiment 7.2 will be lost when the Capstone is closed!

3. Close PASCO Capstone by clicking the red-cross at the upper corner, then select "discard".

4. Turn off PASCO 850 interface and shut down the computer.

5. Clean up your bench.

Lab report on Experiment 7.2

1. Your lab report should be in the required format described in the "Introduction" of the lab manual.

2. Graph Page #1 (similar to Fig. 4) and Graph Page #3 (similar to Fig. 5) should be included in your lab report.

3. Tables 1 and 2 in Experiment 7.2 should be included in your lab report.

4. It is required that the answers or solutions to the questions (see below) should be included in your lab report.

5. You can tear those pages out of the lab manual as a part of your lab report, which contain measured (raw) data and analyzed data, answers to questions. The data sheets must be checked and signed by your lab TA.

Questions

1. Suppose the inner and outer solenoids are used as the primary and secondary coils respectively, and the inner solenoid is connected to the Output Ports of the signal generator on 850 interface as shown in figure 2. If the signal amplitude from the signal generator on 850 interface is set to 10 V, the 850 interface will be crashed. Explain the cause of the crash (hint: resistances of the inner and outer coils are 0.2 and 77 Ω respectively).

2. Can a transformer convert dc voltage? Why?

3. Which of the two measured values of V_s / V_p in Table 1 is close to the ratio N_s / N_p ? Why?

4. An electric doorbell requires a 12 V, 60 Hz ac supply.
   (a) What turn ratio is required on a transformer to operate from a 120 V, 60 Hz ac supply?

(b) Is this a step – up or step – down transformer?

(c) If the transformer is connected the wrong way around the circuit, what voltage is supplied to the doorbell?

Experiment 8.1
Series LCR resonance

Purpose
Study resonance in a series inductor-capacitor-resistor (LCR) circuit by examining the voltage across the resistor as a function of frequency of the applied sine wave.

Equipment
L-C-R board with L = 3.3 mH, C = 0.39 μF, R = 100 Ω, PASCO 850 interface, one voltage sensor (PASCO UI-5100), computer.

Theory
1. Series LCR resonance
The amplitude of the ac current (I_0) in a series LCR circuit (shown in Fig. 1) depends on the amplitude of the applied voltage (V_0) and the impedance (Z) of the LCR circuit. Notice that the upper case I_0, V_0, V_R, V_L, V_C represent the amplitudes of the current and voltages, the lower case $i, \upsilon, \upsilon_R, \upsilon_L, \upsilon_C$ represent the instantaneous values of the current and voltages.

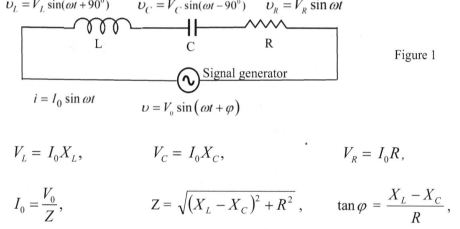

$$\upsilon_L = V_L \sin(\omega t + 90^0) \qquad \upsilon_C = V_C \sin(\omega t - 90^0) \qquad \upsilon_R = V_R \sin \omega t$$

L   C   R

Signal generator

Figure 1

$$i = I_0 \sin \omega t \qquad \upsilon = V_0 \sin(\omega t + \varphi)$$

$$V_L = I_0 X_L, \qquad V_C = I_0 X_C, \qquad V_R = I_0 R,$$

$$I_0 = \frac{V_0}{Z}, \qquad Z = \sqrt{(X_L - X_C)^2 + R^2}, \qquad \tan \varphi = \frac{X_L - X_C}{R},$$

where $X_L = \omega L$ (inductive reactance), $X_C = \dfrac{1}{\omega C}$ (capacitive reactance), R (resistance), angular frequency $\omega = 2\pi f$ (f is the linear frequency) and φ is the phase angle between the current (i) and the voltage (υ) of the source.

Since the impedance depends on frequency, the current and voltages vary with frequency. The current will be maximum when the circuit is driven at its resonance frequency:

$$\omega_{res} = 2\pi f_{res} = \frac{1}{\sqrt{LC}} \qquad (1)$$

At resonance, $X_L = X_C$, $Z = Z_{min} = R$, $I_0 = I_{max} = V_0 / R$, $\varphi = 0$, i.e., the current $i(t)$ and voltage $\upsilon(t)$ are in phase, and $V_R = V_{R\max} = V_0$. These results are derived based on an assumption: the inductor is ideal which has only inductive reactance without resistance. In practice, any real inductor has resistance and therefore at resonance $V_R = V_{Rmax} < V_0$, however, this fact won't affect the present experiment.

2. Lissajous curve

If we apply the source voltage $\upsilon(t) = V_0 \sin(\omega t + \varphi)$ and the voltage across the resistor $\upsilon_R(t) = V_R \sin \omega t$ to the y-axis and x-axis inputs respectively of an oscilloscope, the phase difference between υ and υ_R will produce an elliptical pattern on the screen, which is a Lissajous curve. Figure 3 shows some Lissajous curves and the phase angles φ. At resonance of the series LCR circuit, $\varphi = 0$, Lissajous curve becomes a straight line which is the most intuitive and accurate approach for finding the resonance frequency of the series LCR circuit.

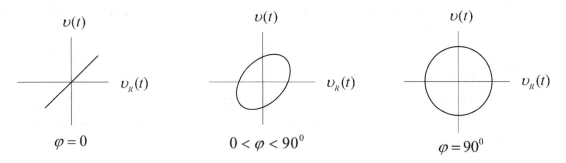

Figure 2   Lissajous curves

3. Fast Fourier transform (FFT)

A fast Fourier transform (FFT) is an algorithm to compute the discrete Fourier transform (DFT) and its inverse. Fourier analysis converts time (or space) to frequency and vice versa. FFT has been described as "the most important numerical algorithm[s] of our lifetime, and is widely used for many applications in engineering, science, and mathematics.

A sine wave consists of a single frequency only, and its spectrum in frequency domain is a single point. Theoretically, a sine wave exists over infinite time and never changes. The mathematical transform that converts the time domain waveform into the frequency domain is called the Fourier transform, and it compresses all the information in the sine wave over infinite time into one point. The fact that the peak in the spectrum in frequency domain shown in Fig. 3 has a finite width is an artifact of the FFT analysis.

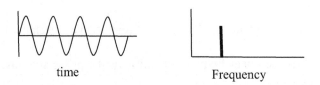

Figure 3   Sine wave spectrum in time and frequency domains

4. Characteristics of series LCR resonance

Figure 4 shows three characteristics of a series LCR resonance (at $f = f_{res}$).

(a) Lissajous curve of $\upsilon(t)$ vs $\upsilon_R(t)$ is a straight line (Fig. 4A).

(b) $\upsilon(t)$ and $\upsilon_R(t)$ curves line up (in phase) as shown in Fig. 4B.

(c) FFT gives the accurate value of the resonance frequency at which V_R reaches its maximum value $V_{R\max}$ (Fig. 4C). Notice that the $\upsilon_R(t)$ curve is enlarged by a factor of 2.

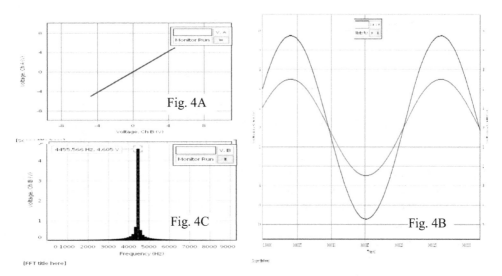

Figure 4   Three characteristics of series LCR resonance

Preparation

The sketch of PASCO 850 interface is given in the "Introduction" of this lab manual.

Caution: Turn off the power of PASCO 850 interface before you make connections. Don't let the two output terminals from the Output Ports on 850 interface touch each other anytime.

1. Connect the circuit as shown in Fig. 5
    (a) Use one red and one black banana leads to connect the LCR series and the Output Ports (banana jacks) for signal generator on 850 Interface as shown in Fig. 5.

    (b) Use a voltage sensor to connect the resistor R and Port A of the Analog Inputs on 850 Interface.

    Attention: the two black leads must be connected to the R-end of the LCR series shown in Fig. 5.

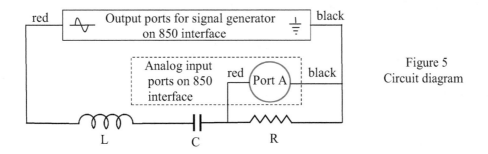

Figure 5
Circuit diagram

2. Turn on PASCO 850 Interface and computer. Log in computer using Password: phy2049.

3. Start PASCO Capstone by double clicking the symbol on computer. The Workbook Page appears.

4. Hardware Setup on the Workbook Page
   (a) In the Tools Palette, click on the "Hardware Setup" icon, the Hardware Setup panel appears with a picture of the PASCO 850 Interface along with a 'voltage sensor' icons on the top of Port A on the Analog Inputs.
   Attention: If the 'voltage sensor' icon does not show up automatically, click Port A on the Analog Inputs (in the PASCO 850 Interface picture), a drop down menu of sensors appears. Type in "v", in the pop-up select "voltage sensor", then a 'voltage sensor' icon appears.
   (b) Click the Output Ports on the upper-right corner of the picture of 850 Interface, in the pop-up select the "Output Voltage-Current Sensor", the sensor icon appears on the top.
   (c) Click on "Hardware Setup" icon to close the Hardware Setup panel.
5. Setup Signal Generator on the Workbook Page
   (a) In the Tools Palette, click "Signal Generator" button to open the Signal Generator panel.
   (b) Click on "850 Output 1" and make the following selections:
   - Wave form: Sine
   - Sweep: Single
   - Amplitude: 5 V
   - Initial frequency: 1000 Hz
   - Final frequency: 10000 Hz
   - Duration time: 20 s
   - Step frequency: 0.1 Hz
   - Auto
   (c) Click "Signal Generator" button to close it.
6. Setup scope display Page #1 on the Workbook page:
   (a) Double click the scope icon in the Displays palette, a scope Page #1 appears on Workbook page. **Note:** the left side of the scope Page #1 is blocked by the Signal Generator panel. Click on the red pin at the upper-right corner of the Signal Generator panel, the full scope Page #1 is displayed on the right side of the Signal Generator panel.
   (b) Click on the y-axis label and in the pop-up select "Output Voltage, Ch 01 (V)" --- the output voltage $\upsilon(t)$ of the sine wave from the Signal Generator.
   (c) Click on the x-axis label and in the pop-up select "Ch A (V)" --- the voltage υ_R across the resistor.

Data collection: Observe Lissajous curve of υ vs υ_R to judge the occurrence of the resonance

1. In the Controls Palette (at the bottom of the Workbook Page) select "Fast Mode" and set sampling rate to 500 kHz.
2. Click the "Monitor" button at the left end of the Controls palette to start data collection. You may not able to see the whole ellipse because of the x-axis scale is too small. Click "Stop" to end the data collection. Click and drag the horizontal cursor ⇔ from right to left to increase the x-axis scale until you see the whole ellipse in the middle of the scope display.
3. Click the "Monitor" button to start data collection again, you can now see the evolution from ellipse → line → ellipse.
4. Just before the ellipse becomes a straight line click "Stop" button to end the data collection. Your goal is to obtain a straight line which is one of the three characteristics of the series LCR resonance.
5. You may need to repeat steps 3 and 4 several times to obtain a good straight line after "Stop" button is clicked.
   Attention: Make sure that the Lissajous curve of $\upsilon(t)$ vs $\upsilon_R(t)$ must be a straight line, otherwise, the $\upsilon(t)$ and $\upsilon_R(t)$ curves won't line up implying that you do not find the accurate f_{res}.
6. If the Lissajous curve is a straight line, set "Amplitude: 0V", then click the "Off" button in the Signal Generator panel to turn off the power of the Signal Generator.

Data analysis:

1. Check the alignment of the $v(t)$ and $v_R(t)$ curves

   (a) Double click the "Graph" icon in the Displays palette, a graph display window appears on Page #1.
   - Click on the y-axis label and in the pop-up select "Output Voltage Ch 01 (V)".
   - Click "Add new y-axis to active plot area" on the top of the graph Tool bar, the new y-axis is displayed on the right side of the second scope display.
   - Set the new y-axis to Voltage, Ch A (V). Now both $v(t)$ and $v_R(t)$ curves are displayed.
   - Change the x-axis scale by dragging the horizontal cursor $\Leftrightarrow$ from left to right until only two or three cycles are displayed from which it is easier to check the alignment of the $v(t)$ and $v_R(t)$ curves.

   (b) If the peaks of the $v(t)$ and $v_R(t)$ curves do not line up well, you need to re-do the data collection and make sure that the Lissajous curve is a good straight line, then re-do this data analysis.

2. FFT analysis to determine exact resonance frequency
   Double click the FFT icon in the Displays palette to open FFT display.
   (a) Click on the y-axis label and in the pop-up select "Voltage Ch A (V)". A FFT curve appears.
   (b) Set the x-axis scale to 0 --- 10000 Hz.
   (c) Adjust the y-axis scale so that the peak of the FFT curve is not too high.
   (d) Click "auto adjust sample rate" (the second icon in the FFT graph Toolbar) to make this function inactive.
   (e) To increase the frequency resolution of the FFT curve, right click "increase number of bins" icon in the FFT graph Toolbar and in the pop-up select "properties", then click on FFT, in the row of "Maximum Number of Bins" select 4096 and click OK.
   (f) Click the Coordinates Tool (called Smart Cursor) ($\cdot\overset{\cdot}{\cdot}\cdot$) from the FFT graph Toolbar, a square box appears, click on the box and drag it until it is directly above the peak of the FFT curve (above it, not on the line – you should see an arrow pointing down to the peak). Release the Smart Cursor and it will snap to the peak. The two numbers in the info box are [frequency (*f*), voltage (*V*)]. The value of the frequency is just the resonance frequency f_{res} and the value of voltage is the $V_{R\max}$.

3. Record the values of f_{res} (exp) and f_{res} (*theory*) in Table 1.

   Table 1 Comparison of f_{res} (exp) with f_{res} (*theory*)

f_{res}(exp)	f_{res}(*theory*)	% error

4. Setup a Text Box on Page #1:
   (a) Double click the Text Box icon in the Displays palette to open it on Page #1.
   (b) You may reduce the size of the Text Box by dragging an arrow cursor ($\Leftrightarrow$) along the diagonal from a corner of the blue frame.
   (c) You can move the Text Box around by clicking on and dragging one of the four sides (blue frame) to where you want to place it.
   (d) Type in content similar to that in Fig. 6.
   When you put mouth cursor at any point in the Text Box a command ribbon appears at the top of Text Box. You can change the font and font size by click "A" and in the pop-up select your favor font and font size.

   Your Graph Page #1 should be similar to Fig. 6.

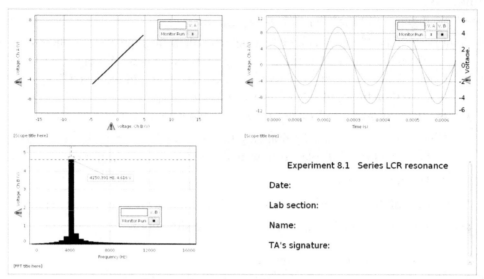

Figure 6 Data of Experiment 8.1

Work to be done:

1. Print Graph Page #1 (similar to Fig. 6) as a part of your lab report on Experiment 8.1.
   Click on "File" at the top-left" corner, in the pop-up select "Print Page Setup" and "Landscape", then print it out.
2. Let your TA check your Graph Page #1, if it is OK your TA will sign it.
   Caution: Make sure that your printed Graph Page #1 is OK before you close the Capstone. All the data collected in Experiment 8.1 will be lost when the Capstone is closed!
3. Close PASCO Capstone by clicking the red-cross at the upper corner, then select "discard".
4. Turn off the 850 interface, leave the computer on.
5. Clean up your bench and setup apparatus of Experiment 8.2.

Experiment 8.2
Phase angle φ versus frequency f

Purpose
Measure phase angle φ of the series LCR circuit at $f < f_{res}$ (theory), $f = f_{res}$ (theory), $f > f_{res}$ (theory).

Equipment
L-C-R board with L = 3.3 mH, C = 0.39 μF, R = 100 Ω, PASCO 850 interface, one voltage sensor (PASCO UI-5100), computer.

Theory   (same as in Experiment 8.1)

Preparation
Caution: **Turn off the power of PASCO 850 interface before you make connections. Don't let the two output terminals from the output ports on 850 interface touch each other anytime.**
1. Connect the circuit as shown in Fig. 5 of Experiment 8.1.
2. Turn on PASCO 850 Interface.
3. Start PASCO Capstone by double clicking the symbol on computer.
    The Workbook Page appears.
4. Hardware Setup on the Workbook Page
    Repeat steps in "Hardware Setup" in Experiment 8.1 with the following changes:
    In the Controls Palette (at the bottom of the Workbook Page)
    (a) select "Continuous Mode",
    (b) set sampling rate to 500 kHz,
    (c) click "Recording conditions", select "Stop condition"/"Time base" and set to 0.02 s.

Measurement #1:   Phase angle φ_{exp} at $f = 1000\,\text{Hz} < f_{res}$ (theory)
1. Setup graph Page #1 on the Workbook page:
    (a) Double click the graph icon in the Displays palette to open a graph on Page #1.
    (b) Click on the y-axis label and in the pop-up select "Output Voltage, Ch 01 (V)" --- the voltage $\upsilon(t)$ of the sine wave from the Signal Generator.
    (c) Click "Add new y-axis to active plot area" on the top of the graph Tool bar, the new y-axis is displayed on the right side of graph Page #1.
        Set the new y-axis to Ch A (V) --- the voltage $\upsilon_R(t)$.
2. Setup Signal Generator on the Workbook Page
    (a) In the Tools Palette, click "Signal Generator" button to open the Signal Generator panel.
    (b) Click on "850 Output 1" and make the following selections:
        *   Wave form: Sine
        *   Sweep: Off
        *   Frequency: 1000 Hz
        *   Amplitude: 5 V
        *   Auto
    (c) Click "Signal Generator" to close it.

3. Data collection
   (a) Click the Record button at the left end of the Controls palette to start data collection which will be automatically stopped at 0.02 s. $\upsilon(t)$ and $\upsilon_R(t)$ curves are displayed.
   (b) Click on the "Data Summary" button at the left edge of the page, right click on Run #1 and rename it as "1000 Hz".

Measurement #2: Phase angle φ_{exp} at $f = f_{res}$ (theory)

Repeat all the three steps in **Measurement #1** but with the following two changes:
1. In step 2, in the Signal Generator panel change the Frequency from 1000 Hz to f_{res} (theory).
2. Rename Run #N as " f_{res} (theory)".

Measurement #3: Phase angle φ_{exp} at $f = 8000\,\text{Hz} > f_{res}$ (theory)

Repeat all the three steps in **Measurement #1** but with the following two changes:
1. In step 2, in the Signal Generator panel change the Frequency from f_{res} (theory) to 8000 Hz.
2. Rename Run #N as "8000 Hz".

When all the three measurements are done, set "Amplitude: 0 V", then click the "Off" button in the Signal Generator panel to turn off the power of the Signal Generator.

Data analysis on graph Page #1

1. Add two graphs on Page #1 to display all data from Measurements #1, #2 and #3.
   Repeat step 1 in Measurement #1 to make total three graphs on Page #1 with the same x- and y-axis labels.
2. On each graph click the Data Selection Tool (⬆▼) so that it is pushed out, click on the small black triangle then select one of the three frequencies ("1000 Hz", " f_{res} (theory)" , and "8000 Hz") on one of the three graphs respectively.
3. Change the x-axis scale by dragging the horizontal cursor ⇔ left to right until only one or two cycles of sine wave are displayed.
4. Label the peaks:
   (a) On each graph use the Coordinates Tool (called Smart Cursor) (⌖) to label the adjacent two peaks of the $\upsilon(t)$ and $\upsilon_R(t)$ curves.
   (b) The two numbers in the info box are [time (t), voltage (V)]. The time difference between the two labeled peaks will be used to calculate the phase angle φ_{exp}.
   (c) Change the significant number for time (t) in the info box.
      You may get $t(\upsilon_R) - t(\upsilon) \approx 0$ from above peak labeling results. To avoid this situation, you need to increase the significant number of time (t) in the info box:
      • Right click the middle point of the info box and in the pop-up select "Tool property".
      • Click "Numerical format" and select "Horizontal coordinator".
      • Check the box of "override default number format".
      • In the row of "Number Style", select "Significant Figures".
      • In the row of "Number of significant number", select 4.
      • Click OK.

5. Record $t(\upsilon_R)$ and $t(\upsilon)$ in Table 1. **Note: In Fig. 7 the black and blue curves correspond to $\upsilon_R(t)$ and $\upsilon(t)$ curves respectively.**

Table 1 $t(\upsilon_R)$ and $t(\upsilon)$ values at the adjacent two peaks of $\upsilon(t)$ and $\upsilon_R(t)$ curves.

	$f = 1000\,Hz$	$f = f_{res}$ (theory)	$f = 8000\,Hz$
$t(\upsilon_R)$ (s)			
$t(\upsilon)$ (s)			

6. Calculate φ_{exp} and φ_{theory} at the three frequencies listed in Table 3. The details of your calculations on φ_{exp} and φ_{theory} will be the answers to question 1 in the section of "**Questions and exercises**".

Table 2 Comparison of φ_{exp} with φ_{theory}

	$f = 1000\,Hz$	$f = f_{res}$ (theory)	$f = 8000\,Hz$
$\Delta t = t(\upsilon_R) - t(\upsilon) =$			
$\varphi_{exp} = \omega\,\Delta t \ = \ 2\pi f\,\Delta t$			
$\varphi_{theory} = \tan^{-1}\left(\dfrac{X_L - X_C}{R}\right)$			

7. Create a Text Box on Page #1 and type in content similar to that in Fig. 7.

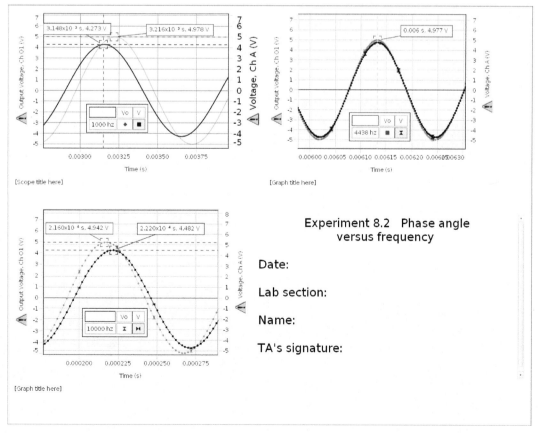

Figure 7 Data of Experiment 8.2

Work to be done:
1. Print Graph Page #1 (similar to Fig. 7) as a part of your lab report on Experiment 8.2.
   Click on "File" at the top-left" corner, in the pop-up select "Print Page Setup" and "Landscape", then print it out.
   Let your TA check your Page #1, if it is OK your TA will sign it.
   Caution: Make sure that your printed Page #1 is OK before you close the Capstone. All the data collected in Experiment 8.2 will be lost when the Capstone is closed!
2. Close PASCO Capstone by clicking the red-cross at the upper corner, then select "discard".
3. Turn off PASCO 850 interface and shut down the computer.
4. Clean up your bench.

Lab report on Experiments 8.1 and 8.2
1. Your lab report should be in the required format described in the "Introduction" of the lab manual.
2. Two Graph Pages similar to Figs. 6 and 7 should be included in your lab report.
3. Table 1 in Experiment 8.1, Tables 1 and 2 in Experiment 8.2 (including the required formulas) should be included in your lab report.
4. It is required that the answers or solutions to the questions (see below) should be included in your lab report.
5. You can tear those pages out of the lab manual as a part of your lab report, which contain measured (raw) data and analyzed data, answers to questions. The data sheets must be checked and signed by your lab TA.

Questions and exrcises:
1. Show the details of your calculations on φ_{exp} and φ_{theory} (Table 2).

2. Based on the calculated phase angles φ_{theory} or φ_{exp} in Table 2, draw a phasor diagram for each of the two frequencies: $f = 1000\,Hz$ and $8000\,Hz$, and complete the following two statements.
    (a) At $f =$ _____ Hz, the voltage (v) across the signal generator leads the current (i).
    (b) At $f =$ _____ Hz, the voltage (v) across the signal generator lags the current (i).

3. Phase angle φ is the phase difference between the current $i(t)$ and the voltage $v(t)$ of the sine wave source. In Experiment 8.2 you actually measure the phase angle φ between $v_R(t)$ and $v(t)$ rather than $i(t)$ and $v(t)$, why ?

4.   In Fig. 5, can you put R between L and C for performing Experiments 8.1 and 8.2? Why?

Experiment 9
Reflection, refraction and total internal reflection

Purpose
1. Study physics laws of reflection, refraction and total internal reflection.
2. Measure focal lengths of convex and concave mirrors.
3. Measure index of refraction of water.

Equipment
Ray optics kit (OS-8516), light source (OS-8517) with power supply, ruler (30 cm), white papers, compass, protractor, refraction tank (RT-100N), reading light.
Four demonstrations (see figures on page 73): Miracle mirror, Illusion scope, Total internal reflection: laser beam travels through fiber optic cable, Carnival (laughing) mirror.

Theory
Most objects reflect a certain portion of the light falling on them. Suppose a ray of light is incident on a flat surface shown in Fig. 1, the angle of incidence θ_a is the angle that the incident ray makes with respect to the surface normal, a line drawn perpendicular to the surface at the point of incidence. The angle of reflection θ_r is the angle that the reflected ray makes with the surface normal. The Law of reflection describes the behavior of the incident and reflected rays.

Law of Reflection
The incident ray, the reflected ray, and the surface normal all lie in the same plane, called the plane of incidence, and the angle of reflection θ_r equals the angle of incidence θ_a .

$$\theta_a = \theta_r \qquad\qquad (1)$$

When light strikes the interface between two transparent materials, such as air and water, the light generally divides into two parts, as Fig. 1 illustrates. Part of the light is reflected, with the angle of reflection equaling the angle of incidence. The remainder is transmitted across the interface. If the incident ray does not strike the interface at normal incidence, the transmitted ray has a different direction than the incident ray. The ray that enters the second material is said to be refracted.

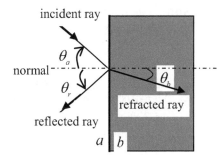

Figure 1

Snell's Law of Refraction
When light travels from a material with refractive index n_a into a material with refractive index n_b, the refracted ray, the incident ray, and the surface normal to the interface between the materials all lie in the same plane. The angle of refraction θ_b, is related to the angle of incidence θ_a by

$$n_a \sin\theta_a = n_b \sin\theta_b \qquad\qquad (2)$$

Equation (2) is known as Snell's law.

Total Internal Reflection

When light passes from a medium of large refractive index into one of small refractive index --- for example, from water to air --- the refracted ray bends away from the surface normal, as shown in Fig. 2. As the angle of incidence increases, the angle of refraction also increases. When the angle of incidence reaches a certain value, called the critical angle, θ_c, the angle of refraction is 90^0 (ray 3 in Fig. 2). When the angle of the incidence exceeds the critical angle (ray 4 in Fig. 2), there is no refracted light. All incident light is reflected back into the medium from which it came, a phenomenon called total internal reflection. Total internal reflection occurs only when light travels from higher-index medium toward lower-index medium. It does not occur when light propagates in the reverse direction --- for example, from air to water.

The critical angle θ_c can be determined from the Snell's Law since $\theta_a = \theta_c$ and $\theta_b = 90^0$ (ray 3 in Fig. 2).

$$\sin \theta_C = \frac{n_b \sin 90^0}{n_a} = \frac{n_b}{n_a}, \quad n_a > n_b \qquad (3)$$

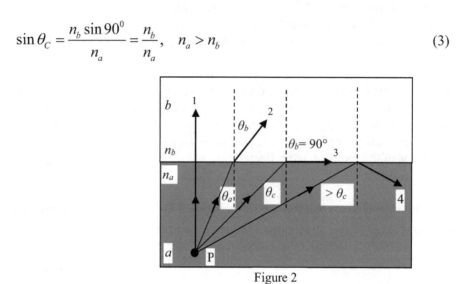

Figure 2

Note: The light source (OS-8517) can be used as ray box. To select the number (1, 3, 5) of white rays, slide the plastic mask, which is fastened to the front of the box until you see the desired number of rays.

Measurement #1: Reflection

1. **Measure the angle of incidence and the angle of reflection at the plane interface**
   (a) Place a sheet of white paper on lab table and place light source (OS-8517) on the paper. Adjust the slit mask in the front of light source until only one light ray is shining.
   (b) Place a triangular mirror on the paper with its plane surface facing the light source as shown in Fig. 3.

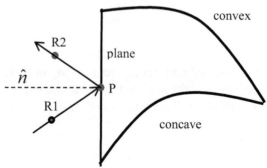

Figure 3   Triangular mirror with convex, concave and plane reflective surface

(c) Rotate the mirror until the angle between the incident and the reflected rays is ~ 60°.

(d) Trace the plane surface of the triangular mirror by drawing a straight line along the bottom edge of the plane surface. Mark points R1, R2 on the incident and reflected rays respectively and the incident point P on the traced line of the plane surface as shown in Fig. 3. The straight line through points R1 and P gives the incident ray, while the straight line through points P and R2 gives the reflected ray. Draw a surface normal $\hat{n}$ to the plane surface at the incident point P, then measure the angle of incidence and the angle of reflection and record them in Table 1.

(e) On the same white paper, change the angle between the incident and reflected rays from ~ 60° to ~ 90°, then repeat step (d), record the results in Table 1.

Table 1 Measured angle of incidence and angle of reflection at the plane interface

Plane surface	
Angle of incidence (θ_a)	Angle of reflection (θ_r)

2. **Measure the angle of incidence and the angle of reflection at the convex interface**

(a) Place a sheet of white paper on lab table and place light source (OS-8517) on the paper. Adjust the slit mask in the front of light source until only one light ray is shining.

(b) Place a triangular mirror on the paper with its convex surface facing the light source as shown in Fig. 4.

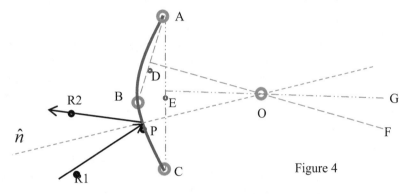

Figure 4

(c) Rotate the mirror until the angle between the incident and the reflected rays is ~ 60°.

(d) Trace the convex surface of the triangular mirror by drawing a curve along the bottom edge of the convex surface. Mark points R1, R2 on the incident and reflected rays respectively and the incident point P on the traced curve of the convex surface as shown in Fig. 4. The straight line through points R1 and P gives the incident ray, while the straight line through points P and R2 gives the reflected ray.

(e) Draw a surface normal $\hat{n}$ to the convex surface at the incident point P using the following steps:
 • Mark three points A, B, C on the trace curve of the convex surface as shown in Fig. 4.
 • Draw straight lines from A to B and from A to C respectively.
 • Mark the middle point D of line AB and the middle point E of line AC.
 • Draw a straight line perpendicular to line AB through point D, i.e., line DF; draw another straight line perpendicular to line AC through point E, i.e., line EG. The two straight lines DF and EG cross at point O which is the center of the spherical convex surface.
 • The straight line through points P and O gives the surface normal $\hat{n}$ to the convex surface at the incident point P as shown in Fig. 4.

(f) Measure the angle of incidence and the angle of reflection and record them in Table 2.

(g) Place another white paper on the table, change the angle between the incident and reflected rays from ~ 60° to ~ 90°, then repeat steps (d), (e), (f). Record the results in Table 2.

Table 2 Measured angle of incidence and angle of reflection at the convex interface

Convex surface	
Angle of incidence (θ_a)	Angle of reflection (θ_r)

3. **Measure the angle of incidence and the angle of reflection at the concave interface**
   (a) Place a sheet of white paper on lab table and place light source (OS-8517) on the paper. Adjust the slit mask in the front of light source until only one light ray is shining.
   (b) Place the triangular mirror on paper with its concave surface facing the light source.
   (c) Use steps similar to steps (c) to (g) in **2.** to measure the angle of incidence and the angle of reflection and record them in Table 3.

Table 3 Measured angle of incidence and angle of reflection at the concave interface

Concave surface	
Angle of incidence (θ_a)	Angle of reflection (θ_r)

Measurement #2: Focal length of a mirror

1. Place a sheet of white paper on lab table and place light source (OS-8517) on the paper. Adjust the slit mask in the front of light source until three parallel light rays are shining.
2. Place the triangular mirror on paper with its concave surface facing the light source.
3. Adjust the mirror so that the middle reflected ray is aligned with the middle incident ray.
4. Trace the concave surface by drawing a curve along the bottom edge of the concave surface.
5. Mark the three incident points on the traced curve of the concave surface. Mark the point where the three reflected rays cross each other which is the focal point of the mirror.
6. Trace the three reflected rays by drawing three straight lines from the focal point to the three incident points.
7. The focal length (*f*) of the concave mirror is obtained by measuring the distance between the middle-incident-point on the surface and the focal point (marked in step 5). Record the focal length in Table 4.
8. Check $f = R / 2$ (*R* is the radius of the concave surface) with a compass using the following steps:
   (a) Extend the straight line from the middle-incident-point on the concave surface to the focal point so that the total length of the straight line is greater than 2*f*.
   (b) Set the separation of the two tips of the compass equal to 2*f*.
   (c) Put the pencil tip of the compass at the middle-incident-point on the concave surface and let the metal tip touch a point on the extension of the straight line.
   (d) Draw a circle centered at the touch point of the metal tip. The match of the drawn circle with the traced curve of the concave surface shows that $f = R / 2$.
9. Steps 1 to 8 can also be used to measure the focal length and the radius of the convex mirror. Note that the reflected rays are diverging and they won't cross on the incident side of the convex mirror. **Attention:** in addition to mark three incident points on the traced curve of the convex surface you need also to mark three points on the reflected rays. Draw three straight lines from the reflected points to the corresponding three incident points. Extend the reflected rays back behind the convex mirror's surface to find the focal point.
   Record the measured *f* and *R* of the convex mirror in Table 4.

Table 4

Concave mirror			Convex mirror		
focal length (f)	radius R	Is $f = R/2$?	focal length (f)	radius R	Is $f = R/2$?

Measurement #3: Refraction of light passing from air into water

A refraction tank is shown in Fig. 5. It consists of a circular bath and light source. The circular bath has a built-in scale enabling the angles of incidence and reflection/refraction to be easily ascertained. The laser light source can be positioned at any point along the circular scale.

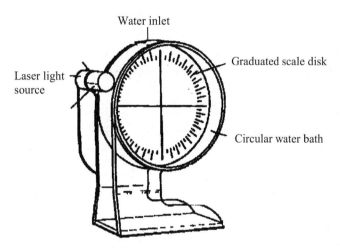

Figure 5   Refraction Tank (RT-100N)

Distilled water is filled up to the line of 90^0-90^0 in the tank, adjust the four screws on the base of the tank to let the water level be parallel to the line of 90^0-90^0. Adjust the laser beam so that the incident point right at the center of the tank as shown in Fig 6.

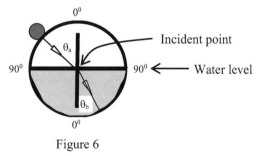

Figure 6

Slowly move the laser light source from its top position (zero on the scale) downward while at the same time observe the path of light. Record θ_a and θ_b in Table 5. Calculate the index (n_{water}) of refraction of water.

Table 5 Measured angle of incidence and angle of refraction at the air to water interface, and n_{water}

Angle of incidence (θ_a)	Angle of refraction (θ_b)	n_{water}	Average of n_{water}

Measurement #4: Refraction of light passing from water into air

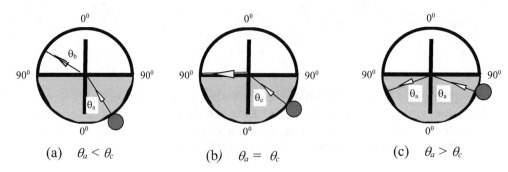

(a) $\theta_a < \theta_c$ (b) $\theta_a = \theta_c$ (c) $\theta_a > \theta_c$

Figure 7

1. Set the angle of incidence as shown in Fig. 7(a). Record the angle of incidence θ_a and the angle of refraction θ_b in Table 6. Calculate the index (n_{water}) of refraction of water.

Table 6 Measured angle of incidence and angle of refraction at the water to air interface, and n_{water}

Angle of incidence (θ_a)	Angle of refraction (θ_b)	n_{water}

2. Set the angle of refraction 90^0, the angle of incidence reaches the critical angle θ_c, as shown in Fig. 7(b). Record θ_c in Table 7. Calculate the index (n_{water}) of refraction of water.

Table 7 Measured critical angle θ_c at the water to air interface, and n_{water}

Angle of incidence ($\theta_a = \theta_c$)	Angle of refraction (θ_b)	n_{water}
	90^0	

3. Set the angle of incidence as shown in Fig. 7(c). All incident light is reflected back into the water, total internal reflection occurs. Record θ_a and θ_r in Table 8.

Table 8 Measured angle of incidence ($\theta_a > \theta_c$) and angle of refraction at the water to air interface

Angle of incidence (θ_a)	Angle of reflection (θ_r)

Demonstrations
1. Miracle mirror (Fig. 8)
    Use a small plastic pig as an object. When the pig is placed on the convex side of the Miracle mirror only one image can be formed: an erect and reduced image. If the pig is placed on the concave side of the Miracle mirror three images can be formed: (a) inverted and reduced image when the pig is placed outside the bowl of the concave surface; (b) inverted and enlarged image when the pig is placed inside the bowl; (c) erect and enlarged image if the pig is placed close to the vertex of the concave surface.
2. Illusion scope (Fig. 9)
    Any item of a suitable size placed in the base appears to float in midair. The 3-D hologram image looks so real but cannot be touched. Notice that the illusion scope consists of two concave mirrors with the same size and same focal length. The vertex of one mirror is the focal point of the other.

3. Total internal reflection: laser beam travels through fiber optic cable (Fig. 10) .

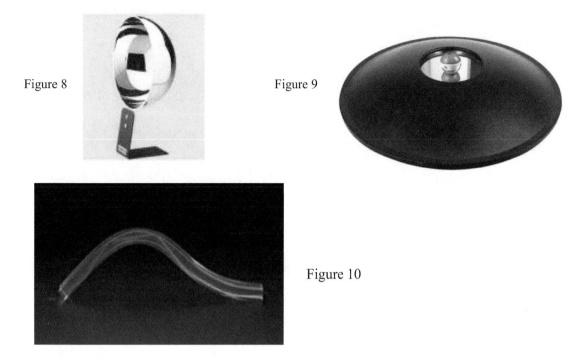

Figure 8

Figure 9

Figure 10

Work to be done:
1. Check that all light sources used in this lab are off, mirrors are placed properly.
2. Let your TA check your data Tables. If they are OK, your TA will sign them.
3. Clean up your bench.

Lab report on Experiments 9
1. Your lab report should be in the required format described in the "Introduction" of the lab manual.
2. Tables 1 to 8 should be included in your lab report.
3. It is required that the answers and solutions to the "Questions and Exercises" (see below) should be included in your lab report.
4. You can tear those pages out of the lab manual as a part of your lab report, which contain measured (raw) data and analyzed data, answers to questions. The data sheets must be checked and signed by your lab TA.

Questions and Exercises
1. Does total internal reflection occur if the refraction of light is passing from air into water? Why?

2. Use Snell's law ($n_a \sin\theta_a = n_b \sin\theta_b$) to explain why θ_b is greater than θ_a in Fig. 2.

3. In Fig. 2, if we increase the index n_b but keep $n_b < n_a$, should the value of θ_c increase or decrease? Give your reasoning.

4. Is the car window mirror concave or convex? Why?

5. The critical angle is 40^0 at the interface from plastic to air. Figure 11 shows a light ray is incident on the top surface of the plastic block. Draw on Fig. 11 the ray path inside the plastic block until it emerges from the block. All the refraction and reflection angles should be drawn to scale using a protractor.

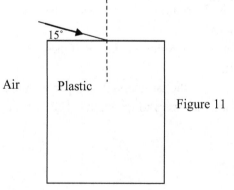

Figure 11

Experiment 10.1
Lenses

Purpose
Determine the focal length of a thin lens and study the image characters.

Equipment
Optics bench (OS-8518), light source (OS-8517) with power supply, converging lens (f = 20cm), screen, ruler (15 cm), reading light.

Theory
A lens of the type shown in Fig. 1 has the property that when a beam of parallel rays passes through the lens (i.e., the object distance $s = \infty$), the rays converge to a point F_2 (i.e., the image distance $s' = f$), (Fig. 1a), and form a real image at that point. Such a lens is called converging lens. Similarly, rays passing through point F_1 $(s = f)$ emerge from the lens as a beam of parallel rays ($s' = \infty$), (Fig. 1b). The points F_1 and F_2 are first and second focal points of the lens respectively. The value of the focal length (f) of converging lens is positive.

Figure 1 The first and second focal points of a converging thin lens. The numerical value of f is positive.

Fig. 2 shows a diverging lens. The beam of parallel rays ($s = \infty$) incident on this lens diverges after refraction (Fig. 2a) and appears to come from the second focal point F_2 which is to the left of the lens, while incident rays converging toward the first focal point F_1 emerge parallel ($s' = \infty$) to the axis, as shown in Fig. 2b. The value of the focal length (f) of diverging lens is negative.

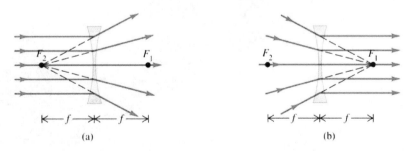

Figure 2 The first and second focal points of a diverging thin lens. The numerical value of f is negative.

The image character, size and position can also be determined by a graphical method called a principal-ray diagram. The rays used in such a diagram shown in Fig. 3 are:

(1) Parallel ray --- a ray parallel to the axis emerges from the lens in a direction that passes through the second focal point F_2 of a converging lens, or appears to come from the second focal point F_2 of a diverging lens;

(2) Central ray --- a ray through the center of the lens is not deviated;

(3) Focal ray --- a ray through first focal point F_1 emerges parallel to the axis.

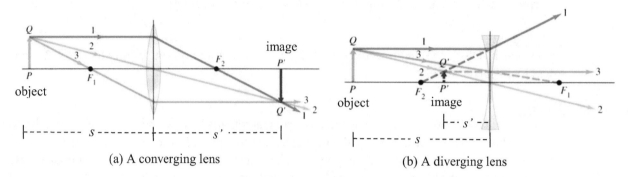

(a) A converging lens   (b) A diverging lens

Figure 3   Principal-ray diagrams showing the graphical method of locating an image formed by a thin lens.

For a thin lens:
$$\frac{1}{f} = \frac{1}{s} + \frac{1}{s'},$$
(1)

where f is focal length, s is the distance between the object and the lens, and s' is the distance between the image and the lens (Fig. 4).

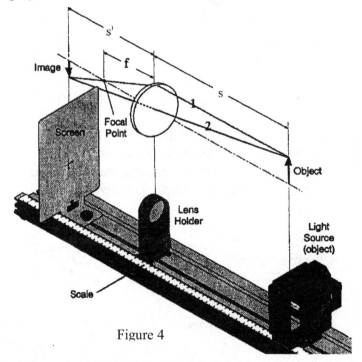

Figure 4

The magnification of a thin lens:
$$m = \frac{h'}{h} = -\frac{s'}{s},$$
(2)

where h' and h are the sizes of the image and the object respectively.

96

Measurement: measure the focal length by plotting $1/s$ vs $1/s'$ (both s and s' vary)

Place the light source at -1 cm on the optics bench. The cross ⊕ attached to the light source is used as an object. Turn on the light source and turn off the room light.

The unit of length is cm for all measured quantities: s, s', h, h' and f.

1. Measure and record the size of the object (h) using a ruler.
2. Place the converging lens to the position as indicated in the Data Table, move the screen until a clear image is on the screen. Measure and record the image position on the optics bench.
3. Measure and record the size of the image (h'). Calculate the magnification $m = h'/h$.
4. Calculate s, s', $1/s$ and $1/s'$.
5. Move the lens to the next position as indicated in the Data Table, repeat the steps 2, 3 and 4.
6. Plot $1/s$ vs $1/s'$ using the data in Table 1 and fit it with a straight line, the intercept gives $1/f$.

Table 1 Data of s, s', h, h', m and f of a converging lens

Object (light source)		Lens	Image					Calculation				f value	
Size (h)	Position on the bench	Position on the bench	Position on the bench	Size (h')	Erect or invert	Real or virtual	$m = h'/h$	s	$1/s$	s'	$1/s'$	fitting result	from manu-facturer
	-1	24											
		25											20 cm
		26											
		28											
		30											
		32											
		35											
		40											
		45											
		50											
		55											
		60											
		65											

Observation:

Place the object at **-1** cm and the screen at **112** cm, move the lens until the clear image is on the screen. Position 1: _____. Position 2: _____.

Experiment 10.2
Telescope

Purpose
Construct a telescope and determine the magnification.

Equipment
Optics bench (OS-8518), two converging lenses (f =10 cm, f =20 cm), screen (graphic paper), ruler (2 m),

Theory
Telescope is used to view large objects at large distances. An astronomical telescope is constructed with two converging lenses: an objective and an eyepiece. The objective lens forms a real reduced image of the object. This image is the object for the eyepiece lens. Objects that are viewed with a telescope are usually so far away from the instrument that the first image is formed very nearly at the second focal point of the objective lens. If the final image formed by the eyepiece is at infinity, the first image must also be at the first focal point of the eyepiece. The distance between objective and eyepiece is therefore the sum of the focal lengths of objective and eyepiece, i.e. $L = f_1 + f_2$ as shown in Fig. 1.

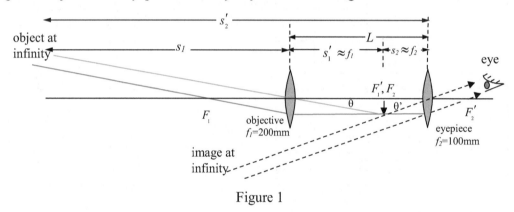

Figure 1

The magnification (M) of a two-lens system is equal to the multiplication of the magnifications of the individual lenses. For $s_1 \rightarrow \infty$ (or $s_1 \gg L$), $s_1 \approx s_2'$,

$$M = M_1 M_2 = \left(\frac{-s_1'}{s_1} \right) \left(\frac{s_2'}{s_2} \right) \approx -\frac{f_1}{f_2} \qquad (1)$$

Procedure
1. Tape the graphic paper (screen) on the wall, the crosshatching on the screen acts as an object.
2. Place the two lenses on the optics bench as shown in Fig. 2.

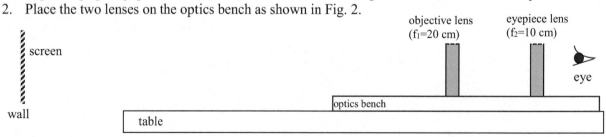

Figure 2

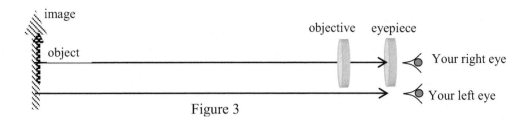

Figure 3

3. Look through the lens with one eye, adjust the position of eyepiece lens until the clear image of the object is on the wall.
4. Open both eyes and look through the lenses at the image with one eye while looking around the edge of the lenses directly at the object with the other eye. Move your head back-and-forth or up-and-down. As you move your head, the image will move relative to the object due to the parallax. To eliminate the parallax, move the eyepiece lens until the image does not move relative to the object when you move your head. Once there is no parallax, the image appears to be stuck to the object (Fig. 3).
   Note: Even there is no parallax, the image may appear to move near the edges of the lens because of lens aberrations.
5. With the parallax eliminated, the virtual image is now in the plane of the object. Record and calculate the positions of the lenses, the object and the image. Find the magnifications M_1, M_2 and M.

Table 2 Measured and calculated data of the 2-lens system shown in Fig. 2

Position of Screen (object) (image)	Position of objective lens (f_1=20cm)	Position Of eyepiece Lens (f_2=10cm)	L	s_1 (cm)	s_1' (cm) (calculated)	s_2' (cm)	s_2 (cm) (calculated)	M_1	M_2	M (M_1M_2)	Invert or erect?

Work to be done:
1. Check that all light sources used in this lab are off. All items used in this lab are placed properly.
2. Let your TA check your data Tables. If they are OK, your TA will sign them.
3. Clean up your bench.

Lab report on Experiments 10.1 and 10.2
1. Your lab report should be in the required format described in the "Introduction" of the lab manual.
2. Tables 1 and 2 as well as the plot and fitting curve in 10.1 should be included in your lab report.
3. It is required that the answers and solutions to the "questions and exercises" should be included in your lab report.
4. You can tear those pages out of the lab manual as a part of your lab report, which contain measured (raw) data and analyzed data, answers to questions. The data sheets must be checked and signed by your lab TA.

Questions and exercises

1. Question on the measurement in **Experiment 10.1**

   (a) Is the image enlarged or reduced when $s > 2f$? Is the image enlarged or reduced when $s < 2f$?

   (b) Is the image erect or inverted? Is the image real or virtual? Why?

2. Question on the observation in **Experiment 10.1**

   Explain why, for a fixed separation between the object and the screen (i.e., $s + s'$ is fixed), there are two positions of the lens at which the images formed on the screen are in focus (clear images).

3. Question on telescope in **Experiment 10.2**

   Hold the optics bench by your hand, look at the tree outside window using your telescope. Is the image of the tree erect or inverted? Is the image of the tree enlarged or reduced?

4. Exercise on 2-lens system used in **Experiment 10.2**

The two converging lenses shown in the figure are identical with focal length (f) of 10 cm. The object height h_o = 3.5 cm. The object distance d_o and the relative positions of the two lenses are given in the figure.

(a) Draw a ray diagram on the figure to show the image position (d_i) and size (h_i) formed by the two lenses.

(b) Describe the characteristics of the image formed by the two lenses:
   - real or virtual?
   - erect or inverted?
   - enlarged, or reduced, or unchanged?

(c) Use the lens equation $\dfrac{1}{d_o} + \dfrac{1}{d_i} = \dfrac{1}{f}$ to verify your findings from your ray diagram.

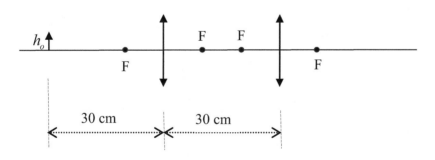

Experiment 11
Dispersion from a prism
and index of refraction

Purpose
1. Study the dispersion spectrum from a prism.
2. Measure index of refraction (*n*) of a prism using the following methods: Snell's law, minimum deviation from a prism, total internal reflection, Brewster's angle and apparent depth.

Equipment
Ray optics kit (OS-8516), light source (OS-8517) with power supply, ruler (15 cm), white papers, protractor, reading light.

Theory of dispersion
A monochromatic (single color or wavelength) light beam in air, obliquely incident on the surface of a transparent medium, is refracted and deviated from its original direction in accordance with Snell's law:

$$n = \frac{c}{\upsilon} = \frac{\sin\theta_1}{\sin\theta_2}, \tag{1}$$

where *n* is the index of refraction of the medium, *c* is the speed of light in vacuum (air), υ is the speed of the light in the medium, and θ_1 and θ_2 are the angles of incidence and refraction respectively.

If the incident beam is not monochromatic, each component wavelength (color) is refracted differently because the wave speed is slightly different for different wavelengths in a medium, e.g. a glass prism (Fig. 1). The dependence of the wave speed (or the index of refraction) on wavelength is called dispersion.

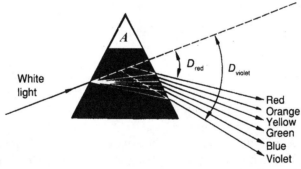

Figure 1   Dispersion. The dispersion light by a glass prism causes white light to be spread out into a spectrum of colors. The angle between the original direction of the beam and an emergent component is called the angle of deviation D for that particular component. The angle *A* is called the prism angle.

The frequency of a wave is unchanged when it propagates from one medium into another. Therefore the greater the wavelength λ the greater the speed of the wave in the medium ($f = \upsilon/\lambda$). Consequently, the indexes (*n*) of refraction are slightly different for different wavelengths.

$$n = \frac{c}{\upsilon} = \frac{f\lambda_0}{f\lambda} = \frac{\lambda_0}{\lambda}, \tag{2}$$

where λ_0 and λ are the wavelengths in vacuum and in the medium respectively.

The dispersion of a beam of white light spreads the transmitted emergent beam into a spectrum of colors, red through violet. The red component has the longest wavelength, so it is deviated least. The angle between the original direction of the beam and an emergent component of the beam is called the angle of deviation (D), and is different for each color or wavelength (Fig. 1).

As the angle of incidence is decreased from a large value, the angle of deviation of the component colors decreases, then increases, and hence goes through an angle of minimum deviation D_m.

The angle of minimum deviation and the prism angle A (Fig. 1) are related to the index (n) of refraction of the prism glass (for a particular color component) through Snell Law by the relationship:

$$n = \frac{\sin\left(\dfrac{A + D_m}{2}\right)}{\sin\left(\dfrac{A}{2}\right)} \tag{3}$$

Measurement #1: **Dispersion spectrum from a prism and measurement of _n_ using the Snell's law**

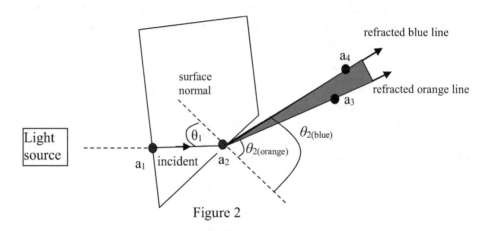

Figure 2

1. Place a sheet of white paper on lab table and place light source (OS-8517) on the paper. Adjust the slit mask in the front of light source until only one light ray is shining.
2. Place the no-shine surface of the rhombus downward to the table. The triangular end of the rhombus is used as a prism in this experiment (Fig. 2).
3. Rotate the rhombus until the angle (θ_2) of the emerging ray is as large as possible and the ray separates into colors. Record what colors are seen, in what order they are and which color is refracted at the largest angle.
4. Trace the outline of the rhombus by drawing straight lines along its bottom edges, mark four points a_1, a_2, a_3 and a_4, as shown in Fig. 2.
5. Remove the rhombus. Draw four straight lines through points (a_1 , a_2), (a_2 , a_3), (a_2 , a_4) , and the surface normal (Fig. 2).
6. Measure the incident angle θ_1, the refracted angles $\theta_{2\,(orange)}$ and $\theta_{2\,(blue)}$ using a protractor. Record the data in Table 1. To make the angle measurements easier you may extend the four straight lines.

Table 1. Data from measurement #1

| incident angle θ_1 | refracted angle θ_2 | | index of refraction n | |
	orange line	blue line	orange line	blue line

Measurement #2: Measure index of refraction (*n*) using the minimum deviation from a prism

The index of refraction of the rhombus can also be determined with the same experimental setup using the minimum deviation from the prism.

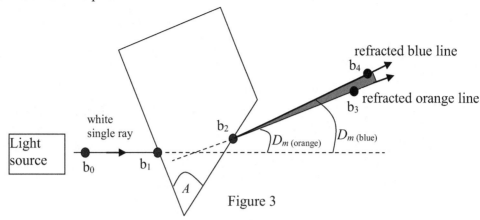

Figure 3

1. Repeat the steps 1 and 2 in **Measurement #1**.
2. Rotate the prism to reduce the angle (θ_2) of the emerging ray to find the position for minimum deviation. Rotate the prism back and forth, and **note the reversal of the direction of motion of spectrum (color rays) when the prism is rotated in one direction**. Stop rotating the prism right at the position of the reversal of motion of the spectrum.
3. Trace the outline of the rhombus, mark five points: b_0, b_1, b_2, b_3 and b_4 as shown in Fig. 3.
4. Remove the rhombus. Draw three straight lines through pints (b_0, b_1), (b_2, b_3), (b_2, b_4) respectively (Fig. 3).
5. Measure $D_{m\,(orange)}$ --- the angle between the lines b_0 b_1 and b_2 b_3. Measure $D_{m\,(blue)}$ --- the angle between the lines b_0 b_1 and b_2 b_4 (Fig. 3). Record the data in Table 2. To make the angle measurements easier you may extend the three straight lines.
6. Calculate the speed of the light υ and the wavelength λ in the rhombus using Eq. (2).
7. Calculate the index of refraction *n* of the rhombus using Eq. (3), the prism angle $A= 45^0$.

Table 2. Data from Measurement #2

	D_m	n (index of refraction)	υ (speed of light in the prism)	λ_0 (wavelength in the vacuum)	λ (wavelength in the prism)
orange line				576 nm	
blue line				436 nm	

Measurement #3: Measure index of refraction (*n*) using the principle of total internal reflection

The index of refraction of the rhombus can also be determined with the same experimental setup using the principle of total internal reflection.
Review the theory of total internal reflection in Experiment 9. Fig. 4 shows the critical total internal reflection at the interface from glass to air.
The index of refraction of the rhombus is *n*.
At the critical incident angle, the Snell's Law gives:

$$n \sin \theta_c = (1) \sin 90^0 \quad or \quad n = \frac{1}{\sin \theta_c} \qquad (4)$$

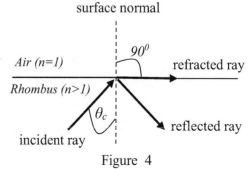

Figure 4

1. Repeat the steps 1 and 2 in **Measurement #1.**
2. Rotate the rhombus until the refracted ray (the ray separates into colors) just barely disappears. The rhombus is correctly positioned if the *red light* has just disappeared as shown in Fig. 5.

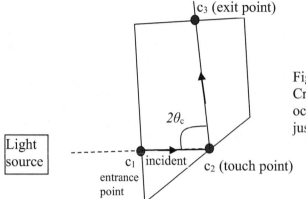

Figure 5
Critical total internal reflection occurs when the refracted ray just barely disappears

3. Trace the outline of the rhombus. Mark three points c_1, c_2 and c_3 (Fig. 5), where c_2 is the touch point at which the refracted red light just barely disappears.
4. Remove the rhombus. Draw two straight lines through points (c_1, c_2) and (c_2, c_3) respectively (Fig. 5).
5. Measure the angle ($2\theta_c$) between the two straight lines using a protractor. To make the angle measurement easier you may extend the two straight lines. Note that the measured angle is twice the critical incident angle because the angle of incidence equals the angle of reflection. Record the data in Table 3.
6. Calculate the index of refraction (n) of the rhombus using Eq. (4).

Table 3. Data from Measurement #3

critical incident angle for total internal reflection $2\theta_c$	index of refraction of the rhombus n

Measurement #4: Measure index of refraction (*n*) using the Brewster's angle
The index of refraction of the rhombus (*n*) can also be determined with the same experimental setup using the Brewster's angle.

At the interface from glass to air shown in Fig. 6 if $\theta_1 + \theta_2 = 90^0$, then the incident angle θ_1 is denoted as θ_B and called the Brewster's angle. Substitution of $\theta_1 = \theta_B$ and $\theta_2 = 90^0 - \theta_B$ into the

Snell's Law $\quad n \sin \theta_1 = (1) \sin \theta_2 \quad$ gives
$\quad n \sin \theta_B = \sin (90^0 - \theta_B) = \cos \theta_B,$

or $\qquad n = \dfrac{1}{\tan \theta_B}$ $\qquad\qquad$ (5)

Further discussion of Brewster's law is given in Experiment 12.

1. Repeat the steps 1 and 2 in **Measurement #1.**
2. Rotate the rhombus until the angle between the refracted and the reflected rays $\beta = 90^0$ as shown in Fig. 6.

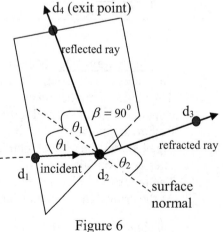

Figure 6

3. Trace the outline of the rhombus. Mark four points d_1, d_2, d_3 and d_4 (Fig. 6).

4. Remove the rhombus. Draw three straight lines through points (d_1, d_2), (d_2, d_3) and (d_2, d_4) respectively. Measure the angle β between the lines of $d_2\,d_3$ and $d_2\,d_4$. If $\beta \neq 90^0$, you need to repeat steps 2, 3 and 4 until $\beta = 90^0$. Then measure the Brewster's angle $\theta_1 = \theta_B$ and record it in Table 4.

5. Calculate n, the index of refraction of the rhombus using Eq. (5).

Table 4. Data from Measurement #4

Brewster's angle ($\theta_1 = \theta_B$)	index of refraction n

Measurement #5: Measure index of refraction (n) using apparent depth

The index of refraction of the rhombus (n) can also be determined using the apparent depth.

Light rays originating from the bottom surface of a block of optical medium refract at the top surface as the rays emerge from the optical medium into the air.

When viewed from above, the apparent depth, d, of the bottom surface of the block is less than the actual thickness, t, of the block. The apparent depth is given by

$$d = \frac{t}{n}, \qquad (6)$$

where n is the index of refraction of the material.

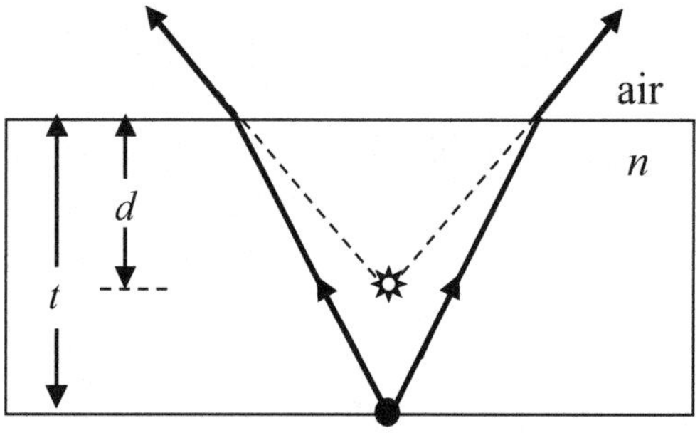

1. Place the light source and the converging lens on a white sheet of paper on the table. Slide the ray mask in the front of the light source until five light rays shine straight into the converging lens. Use a strip of masking tape to block the center three light rays as shown in Fig. 8a.

2. Mark the point "F" (the focal point of the lens) where the two converging light rays cross each other (Fig. 8a).

3. Place the rhombus as shown in Fig. 8b. The surface of the rhombus facing the light source must be exactly at the focal point "F". The cross rays simulate the rays that emerge from the bottom of the rhombus block discussed in the theory (Fig. 7).

Figure 7

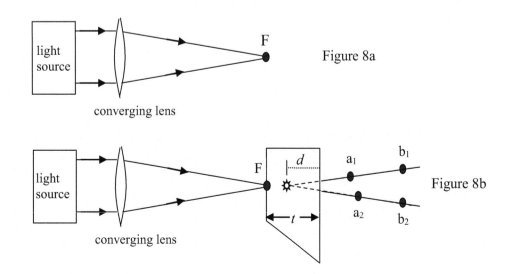

Figure 8a

Figure 8b

4. Trace the outline of the rhombus. Mark the points: a_1, b_1, a_2, b_2 which are on the rays diverging from the surface facing away from the light source as shown in Fig. 8b.
5. Remove the rhombus, turn off the light source, draw two straight lines through points (a_1, b_1) and (a_2, b_2) respectively. The two lines cross at point ✲ inside the outline of the rhombus as shown in Fig. 8b. The distance *d* shown in Fig. 8b is the apparent depth of the rhombus viewed from the right.
6. Measure *t*, the thickness of the rhombus and *d*, the apparent depth of the rhombus viewed from the right. Calculate *n*, the index of refraction of the rhombus using Eq. (6). Record the data in Table 5.

Table 5. Data from Measurement #5

thickness (*t*)	apparent depth (*d*)	index of refraction (*n*)

Work to be done:

1. Check that all light sources used in this lab are off, items used in this lab are placed properly.
2. Let your TA check your data Tables. If they are OK, your TA will sign them.
3. Clean up your bench.

Lab report on Experiments 11

1. Your lab report should be in the required format described in the "Introduction" of the lab manual.
2. Tables 1 to 5 should be included in your lab report.
3. It is required that the answers and solutions to the "questions and exercises" (see below) should be included in your lab report.
4. You can tear those pages out of the lab manual as a part of your lab report, which contain measured (raw) data and analyzed data, answers to questions. The data sheets must be checked and signed by your lab TA.

Questions and Exercises

1. The wavelength of a monochromatic light changes from λ_0 to $\lambda = \lambda_0 / n$ when it propagates from air into an optical medium such as glass. Does the frequency of the light change? Why?

2. Does the index of refraction (n) depend on the color of the light? Explain it with your data.

3. Explain how rainbow is formed.

4. Can the Brewster's angle occur at the interface from air to glass? Why?

5. In **Measurement #3** how does the brightness of the reflected ray change when the incident angle changes from less than θ_c to greater than θ_c? Why?

6. Swimming pool owners know that the pool always looks shallower than it really is. It is important to identify the deep parts conspicuously so that people who can't swim won't jump into water that is over their heads. If a person looks straight down into water that is actually 2.00 m deep, how deep does it appear to be? ($n_{water} = 1.33$).

7. Summarize how many methods can be used to determine the index of refraction (n) of an optical material.

Experiment 12
Polarization of light by absorption and reflection

Purpose
1. Study polarization of light by selective absorption --- Malus' law.
2. Study polarization of light by reflection --- Brewster's law.

Equipment
Optics bench (OS-8518), three polarizers (OS-8520), laser light (OS-8525), rotary motion sensor (PS-2120), high sensitivity light sensor (PS-2176), aperture bracket (OS-8534), rhombus and screen with base, ruler (15 cm), protractor, white papers, reading light, PASCO 850 interface, computer. Two pieces of polarizing films.

Theory
1. Polarization of light
Light, like all electromagnetic radiation, is a transverse wave. The electric field $\vec{E}$ and magnetic field $\vec{B}$ in the light wave are perpendicular to each other, and both are perpendicular to the direction of wave travel as shown in Fig. 1. By convention the "polarization" of light refers to the polarization of the electric field $\vec{E}$. Polarization is a property of waves that $\vec{E}$ can oscillate with more than one orientation.

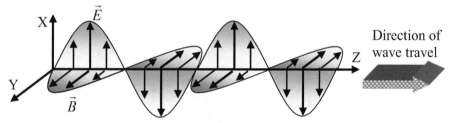

Figure 1 Electromagnetic wave

Light emitted from an ordinary light source is unpolarized, i.e., electric field $\vec{E}$ vectors of the light wave point in all directions perpendicular to the direction of propagation. Figure 2(a) shows different symbols of unpolarized (or natural) light, Fig. 2(b) shows the symbols of linearly polarized (or simply polarized) light. Natural light can be polarized by different means such as selective absorption, reflection, scattering and selective transmission.

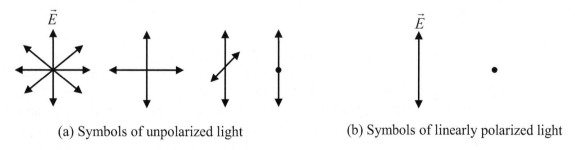

(a) Symbols of unpolarized light   (b) Symbols of linearly polarized light

Figure 2 Symbols of unpolarized and polarized light

2. Polarization of light by selective absorption
Linear polarizer made from dichroic material has a special optical axis called polarizing axis. An ideal polarizer can selectively absorb all the incident light with $\vec{E}$ vector perpendicular to the polarizing axis

and passes all the incident light with $\vec{E}$ vector parallel to the polarizing axis, as a result, the light emerging from the polarizer is linearly polarized. Once the polarized light has been produced, the orientation of its $\vec{E}$ vector can be rotated in the plane perpendicular to the light propagation direction when the light passes through a rotating polarizer. Figure 3 shows the changes in orientation and strength of the electric field $\vec{E}_1$ of the polarized light when it passes through Polarizer #2 and #3 (with polarizing axes $\hat{a}_2$ and $\hat{a}_3$ respectively). Since light intensity is proportional to the square of the electric field strength, when the field strength changes the light intensity will change accordingly.

The following relations can be found from Fig. 3:

(a) $\vec{E}_1 \parallel \hat{a}_1$, $\vec{E}_2 \parallel \hat{a}_2$

(b) $E_2 = E_1 \cos \varphi \quad \rightarrow \quad I_2 = I_1 \cos^2 \varphi$ \hfill (1)

Equation (1) is known as Malus' law which is based on an assumption that all the polarizers are ideal so that they can transmit 100% of the incident light with $\vec{E}$ vector parallel to their polarizing axes and 0% of the incident light with $\vec{E}$ vector perpendicular to their polarizing axes. Real polarizer for visible light usually transmits ~ 80% of the incident light with $\vec{E}$ vector parallel to the polarizing axis and ~ 1% of the incident light with $\vec{E}$ vector perpendicular to the polarizing axis.

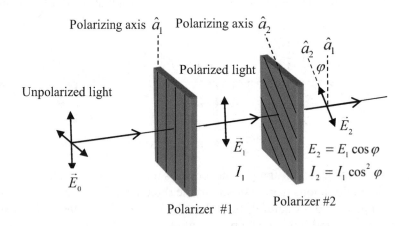

Figure 3 Light passes through two polarizers

3. Polarization of light by reflection

In Fig. 4, unpolarized light is incident on a plane surface between two transparent media, such as air and glass or air and water. The reflected light is usually partially polarized and the degree of polarization depends on the angle of incidence and on the ratio of the indices of refraction of the two materials. It was discovered by Sir David Brewster that when the incident angle is equal to the polarizing angle φ_p (also called Brewster's angle) the reflected light is completely polarized, and in this case the angle of refraction φ_2 becomes the complement of φ_p, i.e., $\varphi_p + \varphi_2 = 90^0$. The electric field of the incident light can be resolved into components: parallel and perpendicular to the plane of incidence. The reflected light is completely polarized with its electric field perpendicular to the plane of incidence.

The Brewster's angle can be related to the indices of refraction of the materials using Snell's law:

$$\tan \varphi_p = n_2 / n_1 \hfill (2)$$

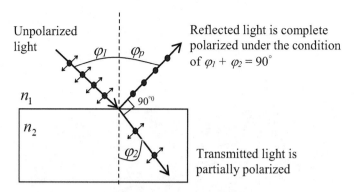

Figure 4 Polarization by reflection at $\varphi_l = \varphi_p$

Observation of polarization of light through two polarizing films:

1. Hold a polarizing film in front of your eye and look at the fluorescent light in the lab. Is there any change in light intensity with and without the polarizing film in front of your eye? Why?

2. Rotate the polarizing film through 360 degrees when you are looking at the fluorescent light. Is there any change in light intensity due to the rotation? Why?

3. Hold two polarizing films together in front of your eye and look at the fluorescent light. Rotate one of the two films through 360 degrees. Is there any change in light intensity due to the rotation? Why?

Measurement #1: Polarization of light by selective absorption through two polarizers
1. Outline
In this measurement laser light (OS-8525, peak wavelength = 650 nm) is used as the light source which is already linearly polarized, so the function of Polarizer #1 is just for making $\vec{E}_1 \parallel \hat{a}_1$ which will serve as a reference line for $\hat{a}_2$ and $\vec{E}_2$ (Fig. 3).

When laser light (with field $\vec{E}_1$) passes through Polarizer #2 which is rotated by hand, the relative intensity of the emerging light is recorded as a function of the angle φ obtained using a Rotary Motion Sensor that is coupled to Polarizer #2 with a drive belt.

This experiment will be performed with the room lights on.

113

Figure 5 Four optical components used in Measurement #1: laser source, Polarizer #1,
Polarizer #2 coupled to Rotary Motion Sensor, Light Sensor with Aperture Disk.

**Caution: Before you make connections turn off PASCO 850 interface and set the power switch of
the laser (OS-8525) to OFF position.**

2. **Apparatus setup**
(a) Mount the following four components on the optics bench as shown in Fig. 5:
- Laser source. Set the power switch to OFF position.
- Polarizer #1. Line up the zero degree mark of the polarizer with the red line drawn at the top
of the polarizer holder and keep Polarizer #1 in this angular position unchanged in
Measurement #1.
- Polarizer #2 coupled to the Rotary Motion Sensor. Line up the zero degree mark of the
polarizer with the red line drawn at the top of the polarizer holder.
- High-Sensitivity Light Sensor coupled to the Aperture Disk

Attention: Push all the components on the optics bench as close together as possible. The Aperture
Disk should not touch Polarizer #2.
(b) Plug the Light Sensor into a PASPORT input on the 850 Universal Interface. Click the low
sensitivity (0-10,000) button on the side of the Light Sensor.
(c) Plug the Rotary Motion Sensor into a PASPORT input on the 850 Universal Interface.
(d) Rotate the Aperture Disk (OS-8534) so that slit #4 is in front of the light sensor.

3. **Software setup**
(a) Turn on PASCO 850 Interface and computer. Log in computer using Password: phy2049
(b) Start PASCO Capstone by double clicking the symbol on computer.
The Workbook Page appears.
(c) Hardware Setup on the Workbook Page
- In the Tools Palette (upper left of the Workbook Page), click "Hardware Setup" icon, the
Hardware Setup panel appears with a picture of the PASCO 850 interface along with a
"Rotary Motion Sensor" icon and a "High Sensitivity Light Sensor" icon below the
PASPORT inputs.

Attention: In case the "Rotary Motion Sensor" icon and/or the "High Sensitivity Light Sensor" icon
do not show up automatically, what you need to do is as follows.

On the picture of the PASCO 850 interface in the Hardware Setup panel click the PASPORT input port where you plug the "Rotary Motion Sensor" into, a drop down menu of sensors appears. Type in a letter r, "Rotary Motion Sensor" appears on the list of the menu, click it, then an icon of "Rotary Motion Sensor" appears below the PASPORT inputs.
Do the same for "High Sensitivity Light Sensor" icon if it does not show up automatically.

- In the Controls Palette (at the bottom of the Workbook Page) select "Continuous Mode" and set sampling rate for Rotary Motion Sensor to 1000 Hz.
- Click on "Hardware Setup" icon to close the Hardware Setup panel.

(d) Data Summary Setup on the Workbook Page #1
- In the Tools Palette, click "Data Summary" icon to open the Data Summary panel. There is an eye-shape icon on the right side of each highlighted row.
- In the highlighted row of "Light Sensor", click on the eye-shape icon, in the pop-up select Relative Intensity and de-select all other options which won't be used in this experiment.
- In the highlighted row of "Rotary Motion Sensor", click on the eye-shape icon, in the pop-up select Angle and de-select all other options.
- Click on "Data Summary" icon to close the Data Summary panel.

(e) Setup Graph Display Page #1
- Double click the "Graph" icon in the Displays palette (top right), a graph display appears.
- Click on the y-axis label and in the pop-up select "Relative Intensity (%)".
- Click on the x-axis label and in the pop-up select "Angle ($^\circ$)".

4. **Align the polarizing axes $\hat{a}_1$ and $\hat{a}_2$ of Polarizers #1 and #2**

(a) Check that the zero degree mark of each polarizer is lined up with the red line drawn at the top of the polarizer holder.

(b) Turn on the laser source, make sure that the laser beam spot is at the center of slit #4 on the Aperture Disk.

(c) Click RECORD and then rotate Polarizer #2 through a full 360 degrees. Click STOP. A curve of relative intensity vs angle appears on the graph with two full minima and maxima. The two maxima are probably not equally high due to the fact that the polarizer is not quite ideal.

(d) From the displayed curve of relative intensity vs angle you know roughly the angle value corresponding to the maximum intensity ($I_{2\max}$ --- the higher of the two maxima). Click RECORD and then rotate Polarizer #2, **stop rotating right at the angle corresponding to $I_{2\max}$** (denoted φ^{start}) and then click STOP. Now the polarizing axes $\hat{a}_1$ and $\hat{a}_2$ of Polarizers #1 and #2 are lined up.

(e) The value of φ^{start} is read from the angle mark at the top of Polarizer #2 which is lined up with the red line drawn at the top of the polarizer holder. Record the value of φ^{start} in Table 1.

(f) Click the white triangle at the lower right of the screen and choose Delete All Runs.

Table 1 Angle φ^{start} at the maximum intensity $I_{2\max}$ through Polarizer #2

Angle φ^{start} at $I_{2\max}$ (degree)

5. **Data collection**

(a) Check that Polarizer #1 is at zero degree position, i.e., its zero degree mark is lined up with the red line drawn at the top of the polarizer holder.

(b) Check that Polarizer #2 is at φ^{start} position.

Attention: To make a good agreement between measured and calculated curves of I vs φ you must start rotating **Polarizer #2 from** φ^{start} **in this data collection.**

(c) Click RECORD and slowly rotate Polarizer #2 (coupled to the Rotary Motion Sensor) through 360 degrees (one revolution), stop rotation when the Relative Intensity is at a maximum and then click STOP. Try to move slowly and steadily through the turning points (i.e., at both the maxima and minima). You may move faster between the turning points. The curve of intensity I_2 vs angle φ (denoted I_2 (exp.)) is now displayed on Graph Page #1.

(d) Click on the "Data Summary" icon. Click on the # of current Run (such as Run #1) and re-label it "Run-P2". Click Data Summary icon to close it.

(e) If the graph does not fill the page, click the Re-size tool (⬜) at the upper left.

(f) You may change the y-axis scale by moving the hand cursor above a number on the y-axis scale, when the hand cursor changes to vertical cursor ⇕ , click and drag the ⇕ cursor along the y-axis to change the y-axis scale so that it is easier to see the minima and maxima.

(g) Click the Smart Cursor (✛) from the graph toolbar. Position the cross-hairs directly at the maximum (or minimum). The two values in the info-box are [Angle ($^\circ$), Intensity (%)].

(h) Record the data listed in Table 2.

(i) Turn off the laser but **keep 850 Interface power on**.

Table 2 Maximum relative intensity $I_{2\max}$ and angle difference $\Delta\varphi$ between two minima

Maximum Relative Intensity $I_{2\max}$ (%)	Angle difference $\Delta\varphi$ between two minima

6. **Calculate I_2 (theory) vs angle φ and compare it with I_2 (exp.)**

Note: Our measurement shows that the polarizers used in this experiment (PASCO OS-8520) transmits ~ 70% (in average) of incident light with $\vec{E}$ vector parallel to the polarizing axis. In order to make a reasonable comparison of I_2(theory) with I_2(exp.) Malus' law in Eq. (1) should be modified as

$$I_2 = I_{2\max} \cos^2 \varphi \qquad (3)$$

Where $I_{2\max}$ is the maximum intensity through Polarizer #2 recorded in Table 2.

(a) In the Tools Palette, click on the "Calculator" icon to open the "Calculator" panel.

(b) In the first row of "Calculations" type in the following formula with units %:

I_2 (theory)=($I_{2\max}$)*(cos([Angle ($^\circ$), ▼]))^2   (4)

Note: replace $I_{2\max}$ in the formula by the measured value of $I_{2\max}$ recorded in Table 2.

(c) Click on ▼ in the formula and in the pop-up select "Run-P2" (named in 5. (b)).

(d) Click "Accept" (above the "Calculations"). If the formula is correct, a message of "Measurement Assignment is OK" and a mathematical equation of the calculation appear in the middle of the Calculator panel.

(e) Click on the "Calculator" icon to close the Calculator panel.

(f) In the graph display click "Add new y-axis to active plot area" on the top of the Graph Tool bar, the new y-axis is added on the right side of the graph window. Click on the new y-axis label and in the pop-up select "I_2 (theory)", the curve of I_2 (theory) vs Angle ($^\circ$) appears on the graph.

(g) The y-axis scale on both left and right sides of the graph should be the same! If not, change the scale of the two y-axes by moving the ⇕ cursor vertically until the same y-axis scale is reached on both sides. Your I_2 (theory) and I_2 (exp.) curves should be similar to that in Fig. 6.

(h) Create a Text Box on Graph Page #1
- Double click the Text Box icon in the Displays palette to open it on Page #1.

- You may reduce the size of the Text Box by dragging an arrow cursor ($\Leftrightarrow$) along the diagonal from a corner of the blue frame.
- You can move the Text Box around by clicking on and dragging one of the four sides (blue frame) to where you want to place it.
- Similarly, you can move and change the size of the Graph Display on Page #1
- Setup the Text Box as shown in Fig. 6 and type in content similar to that in Fig. 6.
- When you move the hand cursor on the Text Box a command ribbon appears at the top of Text Box. You can change the font and font size by click "A" and in the pop-up select your favor font and font size.

7. **Question on Measurement #1**
   (a) Explain why both I_2 (theory) and I_2 (exp.) curves in Fig. 6 have two full minima and maxima.

   (b) How many minima and maxima do you expect in the curve of E_2 vs φ?

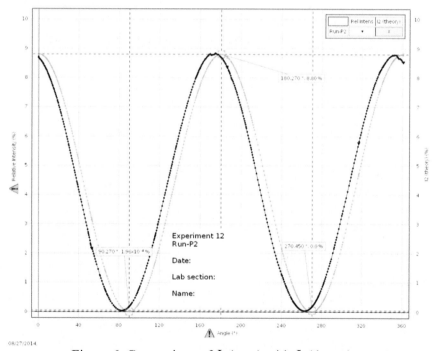

Figure 6 Comparison of I_2 (exp.) with I_2 (theory)

Note: In Fig. 6 the curves of I_2 (theory) and I_2 (exp.) are not coincident because the starting angle for rotating Polarizer #2 is intentionally set to a few degrees off from the angle φ^{start} recorded in Table 1 just for demonstrating the role played by the starting angle φ^{start} .

Work to be done:

1. Print Graph Page #1 (similar to Fig. 6) as a part of your lab report on Experiment 12.
   Click on "File" at the top-left" corner, in the pop-up select "Print Page Setup" and "Landscape", then print it out.
2. Let your TA check your Page #1, if it is OK your TA will sign it.
   Caution: Make sure that your printed Page #1 is OK before you close the Capstone. All the data collected in Experiment 12 will be lost when the Capstone is closed!
3. Close PASCO Capstone by clicking the red-cross at the upper corner, then select "discard".
4. Turn off PASCO 850 interface and shut down the computer.
5. Check that the laser source switch is OFF.
6. Clean up your bench and prepare to perform Measurement #2.

Measurement 2 Polarization of light by Reflection

This experiment is to confirm Brewster's law: the reflected light is completely polarized when the incident angle equals the polarizing angle φ_p in Eq. (2).

1. Place a sheet of white paper on lab table and place light source (OS-8517) on the paper. Adjust the slit mask in the front of light source until only one light ray is shining.
2. Place the rhombus on the paper with the plane surface facing the light source as shown in Fig. 7.
3. Rotate the rhombus until the angle between the reflected and the refracted rays is ~ 90°.
4. Trace the outline of the rhombus by drawing straight lines along its bottom edges, mark four points 1, 2, 3 and 4 as shown in Fig. 7. Move away the rhombus. Draw straight lines from 1 to 2, from 1 to 3 and from 1 to 4. Measure angle ∠ 312 which should be close to 90°. You may need to repeat step 3 several times until the angle ∠ 312 = 90°. Now the reflected light is complete polarized.
5. To verify if the reflected light is complete polarized, place the rhombus back on the tracing paper in the place where ∠ 312 = 90° (the incident ray, the entrance point, the exit point, the reflected ray all match the marks on the tracing paper). Place a screen on the line of the reflection ray (Fig. 7), the reflected ray is shining on the screen.

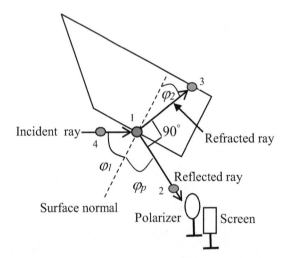

Figure 7 Polarization by reflection

point 1: entrance point
point 2: any point on the reflected ray
point 3: exit point
point 4: any point on the incident ray

6. Place a polarizer in front of the screen (Fig. 7). The light intensity on the screen changes when the polarizer is rotated and the intensity goes to zero (dark) when the polarizer is rotated to a certain position indicating that the reflected ray is complete polarized.
7. Measure the angle of $\varphi_{p\ (measured)}$ on the tracing paper, and record it in Table 3.

8. Using the index of refraction of the rhombus ($n=1.53$), compute the Brewster's angle $\varphi_{p \text{ (calculated)}}$ and record it in Table 3.

Table 3 Measured and calculated angle φ_p

$\varphi_{p \text{ (measured)}}$ (degree)	$\varphi_{p \text{ (calculated)}}$ (degree)

9. **Observation and question**

Rotate rhombus to let angle $\angle 312 \neq 90°$. Repeat steps 5 and 6, you will see that the light intensity on the screen changes with the rotation of the polarizer but never goes to zero. Why?

Work to be done:
1. Let your TA check your three data Tables. If they are OK, your TA will sign them.
2. Check that the power supply used in Measurement #2 is OFF.
3. Place all the items used in this experiment in proper places.
4. Clean up your bench.

Lab report on Experiment 12
1. Your lab report should be in the required format described in the "Introduction" of the lab manual.
2. Graph Page #1 (similar to Fig. 6) should be included in your lab report.
3. Tables 1 to 3 should be included in your lab report.
4. It is required that the answers to the questions in "**Question on Measurement #1**", "**Observation and question**" of Measurement #2, and "**Questions and Exercises**" (see below) should be included in your lab report.
5. You can tear those pages out of the lab manual as a part of your lab report, which contain measured (raw) data and analyzed data, answers to questions. The data sheets must be checked and signed by your lab TA.

Questions and Exercises
1. Derive Eq. (2) using Snell's law.

2. Sunlight reflects off the smooth surface of an unoccupied pool. (a) At what angle of reflection is the light completely polarized? (b) What is the corresponding angle of refraction for the light that is transmitted (refracted) into the water? (c) At night an underwater floodlight is turned on in the pool. Repeat parts (a) and (b) for rays from the floodlight that strike the smooth surface from below.

3. The complete polarization can occur by reflection on the two interfaces: from air to water and from water to air. Can the total internal reflection occur on the two interfaces? Why?

4. Unpolarized light of intensity 3.0 W/m² is incident on two **ideal** polarizing films whose transmission axes make an angle of 60^0 (Fig. 8). What is the intensity of light transmitted by the second film?

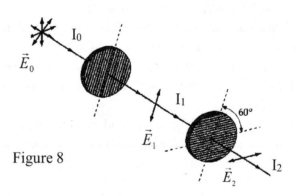

Figure 8